ITALY SINCE 1945

By the same author

CZECHS AND GERMANS

UNDECLARED WAR

ITALY (World Today Series)

THE ROME–BERLIN AXIS

GERMANY'S EASTERN NEIGHBOURS

A GREAT SWISS NEWSPAPER: THE STORY OF THE 'NEUE ZÜRCHER ZEITUNG'

THE KREMLIN SINCE STALIN (a joint translation with Marian Jackson of the book by W. Leonhard)

EUROPE OF THE DICTATORS

THE EUROPE I SAW

FASCISM IN ITALY

ITALY SINCE 1945

Elizabeth Wiskemann

Palgrave Macmillan

First published 1971 by
THE MACMILLAN PRESS LTD
London and Basingstoke
Associated companies in New York Toronto
Dublin Melbourne Johannesburg and Madras

Library of Congress catalog card no. 78–179497

SBN 333 12242 9

ISBN 978-1-349-01123-0   ISBN 978-1-349-01121-6 (eBook)
DOI 10.1007/978-1-349-01121-6

Contents

List of Abbreviations

ACLI Associazione Cristiana di Lavoratori Italiani
AGIP Azienda Generale Italiana di Petroli
CGIL Confederazione Generale Italiana del Lavoro
CISL Confederazione Italiana di Sindacati Liberi
DC Democrazia Cristiana
EFIM Ente per il Finanziamento dell' Industria Manufatturiera
ENEL Ente Nazionale di Energia Elettrica
ENI Ente Nazionale degli Idrocarburi
FIAT Fabbrica Italiana degli Automobili Torino
IRI Istituto per la Ricostruzione Industriale
MSI Movimento Sociale Italiano
PCI Partito Comunista Italiano
PLI Partito Liberale Italiano
PRI Partito Repubblicano Italiano
PSI Partito Socialista Italiano
PSIUP Partito Socialista Italiano d' Unità Proletaria
PSU Partito Socialista Unitario

Publisher's Preface

Miss Elizabeth Wiskemann completed the revision of the typescript of this, her last, book shortly before her death in July 1971.

1 The Republic Emerges

The Italy which emerged from the years of Fascism and war in 1945 was a country of ancient civilisation of many kinds, but one which had been politically united for scarcely seventy-five years. Since the beginning of the twentieth century the north had developed industrially at a quick pace; Turin, the capital city of Piedmont, which had provided Italy's dynasty descended from the Dukes of Savoy, also gave birth to the *Fabbrica Italiana degli Automobili Torino* or FIAT, the great motor industry which was to become one of the country's most precious assets. South of Rome, however, Italy comprised an undeveloped, primitive society where indescribable poverty reigned. North and south did not, indeed, know one another, and since September 1943 the war had cut them apart. In the country as a whole peasants still predominated over industrial workers by about two to one, and by much more than that in southern Italy.

The association of the House of Savoy with the Fascist régime had robbed the dynasty of its prestige, the more so because Fascism had brought the miseries of a lost war together with occupation by a vindictive former ally. Instead the glory of the papacy, spuriously reconciled with the Italian state by Mussolini, was enhanced: since the Concordat of 1929 it laid claim to the special allegiance of Italy as a Catholic confessional state.

Politically, structurally, Italy seemed astonishingly unchanged by the years of Fascism. The corporations vanished overnight, the forbidden trade unions reappeared in the factories in great force, more in evidence than the employers: already in Bari in 1944 they had established the united *Confederazione Generale Italiana del Lavoro* or CGIL to include all organisations of workers. In the person of Ivanoe Bonomi, appointed Prime Minister in June of the same year when the Germans were expelled from Rome, a Premier of the pre-Fascist era returned to office after an interval of nearly twenty-three years, roughly a generation. Until elections could be held six parties were considered to have political reality, the Liberals, the Christian Democrats, the so-called Democracy of Labour grouped around an elderly southern politician called Meuccio Ruini, the Party of Action, the Socialists and their offshoot of January 1921, the Italian Communist Party.

In April 1945 came the final uprising of the Italian partisans in northern Italy; Mussolini was captured and shot, and the German army in Italy surrendered to the Allies. The partisans were able to save vital installations; their Committee of National Liberation, the famous CLN and its subsidiary committees, took over the administration in most northern cities, although until the end of the year the country was occupied by the Allied armies, including the Poles under General Anders. There was a tremendous feeling of national exhilaration. This had been a second *Risorgimento*, the Italians felt, a more real one in which all classes had participated, not only or chiefly the educated as in the nineteenth century. The reunion of the country, after its division since 1943, created an emotional climax; above all a *vento del nord*, a north wind, should now blow away the impure accretions which northerners felt tended to accumulate in the south. There had in fact been no resistance in the south except for a famous four days of fighting – the *Quattro Giornate* – against the Germans in Naples in September 1943 when some very young Neapolitans distinguished themselves.

The fighting resistance in the north had been technically led by General Raffaele Cadorna, the grandson of the General Raffaele Cadorna who conquered Rome from the Pope on 20 September 1870. But the partisans' most beloved leader had been the Piedmontese Ferruccio Parri, 'Maurizio' according to his *nom de guerre*. The Italian Resistance had been strongly influenced by the Communist Party and there had been Catholic partisan groups. But the most inspiring contribution had been made by Parri and his Party of Action with its motto of *Giustizia e Libertà*, the GL people or *Gellisti*. This was the slogan of Carlo Rosselli, a Florentine Jew with Mazzinian links, who had been murdered by Fascist agents in France in 1937. Ten years earlier, in 1927 at Savona, the Fascists had tried Rosselli and Parri who succeeded in flaunting their love of liberty at their judges. After serving his sentence Parri, unlike Rosselli, stayed in Italy. He had friends in the Edison Electricity concern who gave him work, but he never made peace with the Fascists: he was in prison for six months in 1942. He was a romantic figure, marvellously picturesque, and he made a splendid partisan leader. In the last winter of the war the Germans caught him in Milan, but fortunately they handed him over in March 1945, as a pledge of good faith when peace negotiations were begun with the Americans over a truce in Italy. When the country was reunited Parri was the favourite candidate to succeed Bonomi. As heir to Rosselli and leader of the Party of Action, he was on the Left – not quite a Socialist then – and ardently Republican. By an irony of fate Parri was nominated Prime Minister in June 1945 by Prince Umberto, the *Luogotenente* or Regent since the flight and retirement of his father, King Victor Emmanuel III. It was also arranged that a Consultative Assembly or *Consulta* of 429 people nominated among the politically eminent, untarred by Fascism, should meet in September. The *Consulta* elected Count Carlo Sforza to be its President who was as passionately Republican and anti-Savoy as Parri. The winter passed in great economic difficulty and dearth, but in the *Consulta*, which met at Rome in the parliamentary palace of Montecitorio, lively discussion of all kinds of reforms took place. The public showed great interest and crowded into the galleries; it was a novelty to find women among *consultori* and spectators, many of them Communists.

Of the six parties represented in the Government, only the Christian Democrats, the Socialists and the Communists were to find mass support in the local and general elections held in the spring of 1946, the first free elections since 1922. It was remarkable to observe in the voting the survival of local political loyalties which went back to the early years of the century or further. In Milan, for instance, moderate Socialism was again strong. In Emilia and Umbria the old resentment against former papal authority went far to explain the persistent strength of the Communist vote, now and for years to come. In the Marche, for instance at Ancona, the traditional Mazzinian Republican Party was still strong although at this time it represented little but Mazzinian nostalgia. Anti-monarchical feeling had been just as powerful in the Party of Action or among Socialists and Communists. But the Republicans still flourished Mazzini's ivy-leaf;

their faith in the Master was steadfast. They were one of an enormous number of smaller political groups to be found in the ferment of freed Italy.

After the local elections in the spring, on 2 June 1946 a referendum was held on the 'institutional question' – monarchy versus republic. The prophets who announced bitterly that a Fascist monarchy would now be succeeded by a clerical republic were not altogether wrong. Bitterness was increased by the fact that many priests had campaigned for the monarchy, but immediately reversed course when it became clear to them that it would be expedient to capture control of the new republic. The defeat of the monarchy was a defeat for southern Italy which – paradoxically – had voted for the retention of the Piedmontese dynasty. Although the majority for the republic was relatively small – 12,717,923 to 10,719,923 votes – and although its reality was questioned, this controversy died a relatively easy death. There continued to be politicians on the Right who called themselves monarchists for years – sometimes they split into contending factions. Yet they had little but nuisance value, and in spite of the many weaknesses of the republic, the republican form of government ceased to be seriously questioned after a few years. Prince Umberto had gone into exile in Portugal.

On the same day as the referendum a general election was held to elect a Constituent Assembly. 207 Christian Democrats were elected, 115 Socialists and 104 Communists, according to the principles of proportional representation. Of the smaller parties 41 Liberals and 30 *Qualunquisti* were elected. Both terms need explanation. The Italian Liberals claimed political descent from Cavour; to many their liberalism seemed indeed a century out of date. Their enemies said that the only liberty in favour of which they felt strongly was the liberty of the entrepreneur; though not entirely so, they were in effect the party of the less enlightened industrialists and bankers. As for the *Qualunquisti*, they represented a temporary reaction against the reforming spirit of the time. In the outbursts of often brilliant, mostly leftist journalism in Italy in the winter of 1945–6, a third-rate writer by the name of Giannini founded a paper called *L'Uomo Qualunque* ('just any man'). This derided all the idealism of the moment and appealed to rank-and-file former Fascists who were frightened of developments. Giannini's *Qualunquisti* turned out to have no constructive ideas – they could only mock – and they melted away almost as quickly as they had formed into a party.

After the election of 1946 there was no longer any case for six-party government, the *Esarchia* as it had been called. Parri had, not surprisingly, proved incompetent in office, and already in December 1945 the Christian Democrat leader, Alcide De Gasperi, had succeeded him: De Gasperi was to head the governments of Italy for the next seven years. For one more year he felt obliged to keep in office his Marxist colleagues, since the voters had given them so much support. Before examining the tenacious character of De Gasperi it may be well to analyse the personalities of Pietro Nenni and Palmiro Togliatti, his Foreign Minister and his Minister of Mercy and Justice[1] at this time. Already they and the relationship between them profoundly influenced the history of Italy.

The provenance of Italians of their generation was still a cardinal fact; only too often their children and grandchildren were Romans because their parents had been drawn to the capital of a highly centralised state. Togliatti was born in Genoa in 1893; after spending some time in Sardinia, he studied in Turin whose university was a great intellectual centre before, during and after the First World War. At the same time the expansion of Fiat had drawn attention to the new, vast social problems of the age, and the bright young men became involved with these. One of the most prominent was a Sardinian from an impoverished middle-class home who, in spite of ill-health, had managed to win the necessary scholarships to the University of Turin. His name was Antonio Gramsci and he became a passionately enthusiastic champion of the workers, a left-wing Socialist. He believed in self-government in industry, workers' councils, in order to put an end to exploitation. Togliatti followed Gramsci when the latter in January 1921 broke away from the Socialist Party together with Bordiga to help found the *Partito Comunista Italiano or* PCI at a conference at Livorno (Leghorn). Gramsci was a warm, lovable character. Years later in 1927 he was arrested and imprisoned by the Fascist authorities – his letters from prison to his strange Russian wife and their children are famous – and he was released only to die in 1937.

Togliatti, on the other hand, spent the Fascist period in Russia

[1] *Ministro della Grazia e della Giustizia* in Italian.

and survived. When he came back to Italy he at first appeared as an emissary of Stalin. He was cool, even cold, with the air of an exceedingly orthodox cardinal: his faith, too, seemed to have something in common with that of a Catholic prelate, and he seemed intensely aware of the strength of Catholic values in Italy. One would guess that he attached no importance to personal liberty except perhaps his own. He made the impression of an astute, even subtle politician who wanted his own game to succeed; he did not appear to feel much sympathy for suffering; he was not particularly eloquent.

Pietro Nenni could scarcely have been more different. He was born two years earlier at Faenza in the Romagna, a 'countryman' of Mussolini, as people often reminded him. Indeed as young men they had got into trouble and into prison together. Unlike Togliatti, Nenni did not make the impression of an intellectual. He did not think with great clarity, but he was a popular orator of great warmth, a people's tribune; in the Romagna his own people loved to listen to him. Originally he had belonged to the traditional Republican Party in his own turbulent way. By 1922 he had joined the Socialists and in the following year he was made editor of the party newspaper, *Avanti!* He did not, however, count as an important Socialist leader in the pre- and early Fascist period; indeed he worked for a time with Carlo Rosselli, the *Gellista* leader. In exile in France after 1926 he decided that Mussolini had won by dividing the working class and that this must never be allowed to happen again. Hence in France he made his first 'pact of unity' with the Communists in 1934 to which he adhered for over twenty years. It has been part of his fate to be intellectually tripped up by men like Togliatti or by Riccardo Lombardi, an eternal *frondeur* who will appear upon the scene a little later.

Another man with whose fate Nenni's seemed to be entangled was the newly elected President of the Constituent Assembly, Giuseppe Saragat. Born in 1898 in Turin where he too studied, Saragat had been elected at the early age of twenty-seven to the Socialist Party executive with the approval of the veteran leader Filippo Turati. The party had smuggled him out of Italy in 1926 together with another of its leaders, Claudio Treves, at a time when Nenni was left to fend for himself. Saragat was back in Italy before Nenni in 1943. He was always a moderate, who did not trust the Communists not to betray political liberty and sacrifice Italy to Russia. Towards the end of 1943 Saragat had the distinction of being arrested by the Germans, but he was able to escape from them.

Under the fitful chairmanship of Saragat – for he was also constantly in Paris as Italian ambassador there until the end of 1946 – the Constituent Assembly settled down to the work of drawing up a constitution. The first article of the finished document stated that Italy was a democratic republic based on work. A bicameral system was established, with a Chamber of Deputies and a Senate. It was decreed that deputies must be at least twenty-five years of age and senators at least forty; voters for the Chamber must be twenty-one years old and those for the Senate twenty-five. Each deputy was to represent 80,000 voters and each senator 200,000. In addition certain eminent anti-Fascists were nominated for life as senators, for instance Terracini and the Sardinian, Lassu. Article 70 stated that 'Legislative duties are carried out jointly by the two chambers', but there was no provision for resolving a deadlock between them. The method of election, and the duties, of the President of the Republic were defined.

Perhaps the most controversial sections of the constitution were Article 7 and the articles which dealt with the 'Regions, Provinces and Communes' (114–33). The former stated that 'the State and the Catholic Church are, each in its own sphere, independent and sovereign; their relations are regulated by the Lateran Agreements. Such amendments to these Agreements as are accepted by both parties do not require any procedure of constitutional revision.' Like many articles of the new constitution this one was fairly ambiguous; it did, however, incorporate the Lateran Agreements of 1929 which enabled the Vatican to claim obedience from all Italians to its dogmas about marriage and education. There was strong opposition to this virtual establishment of a confessional state from those who inherited the lay tradition of the *Risorgimento*. It was only accepted thanks to Togliatti's startling decision to give it the support of the Communists. Perhaps he already foresaw some kind of Communist–Catholic *rapprochement* such as became possible with the pontificate of John XXIII after 1958. For the moment Togliatti certainly failed to conciliate Pius XII and he shook the allegiance of many of his own followers.

After Article 114 had stated that the republic was to be divided

into regions, provinces and communes, Article 115 laid down
that 'the Regions are constituted as autonomous bodies with their
own powers and functions, according to the principles fixed by
the Constitution'. This at that time shadowy plan to decentralise
the administration of Italy came mainly from the Left, though
the Communists were not at first in favour of it. There was no
doubt that the centralised system which Napoleon had superimposed upon northern Italy, and which had then been trans-
posed from the small kingdom of Sardinia, oddly based on Savoy
and Piedmont, to the kingdom of United Italy had been highly
unsuitable: far from facilitating national unity, it had prevented
the assimilation of southern Italy which detested the Piedmontese
type of prefect sent from Rome to rule the south. For years now
even the Republican Party, which had believed with Mazzini in
a Republic One and Indivisible, had revised its attitude, for it
had adopted the federalism of the Milanese Republican patriot,
Carlo Cattaneo. After all, people said, the old duchies, principali-
ties or republics preserved the essence of Italy in the centuries of
foreign rule; still today an Italian is intensely aware of his region,
Tuscany, say, or Venetia. Unfortunately Article 128 seemed to
make the provinces and communes into rivals of the regions, for
it stated that they too 'are autonomous bodies in conformity with
the principles established by the general laws of the Republic
which determine their function'.

In practice regional autonomy was granted fairly soon, in
accordance with Article 116 of the constitution, to regions with
particular characteristics such as Sicily with its strange and unique
history, Sardinia, the French-speaking Val d'Aosta and the partly
German-speaking Trentino–Alto Adige. It was also planned to
offer it at an early date to Friuli–Venezia Giulia, another mixed-
language area dominated by Trieste. Owing in part to the delay
in the Trieste settlement, regional autonomy for Friuli–Venezia
Giulia was postponed for years; it was finally realised in 1963.

Apart from these outlying regions, each of which constituted
a special case, regional autonomy remained a dead letter until
1970. There were two main reasons for this. Many Italians, in
particular the Liberal Party, but not only Liberals, felt that to
add to the already huge number of officials by creating regional
administrations would be disastrous; partly, too, because it would
be expensive and the Italian Republic never had any money to

spare. There was also anxiety lest no sooner than the unity necessary to national life had been restored, it might be gravely impaired; to re-emphasise the regional emotions of Lombardy or Emilia seemed foolhardy. In particular, as the years passed, it was increasingly feared that the central Italian 'Red Belt' of Emilia, Tuscany, Umbria, would become in effect a separate revolutionary area cutting Italy into at least three pieces. Communist and left-wing Socialists came to predominate in the councils of the communes and provinces here and it seemed that in time they would capture their regional administrations.

In Articles 134-7 of the constitution it was provided that a Constitutional Court should be established; there was also to be a Superior Council of Magistrates and a National Economic and Labour Council. All these took years to materialise, i.e. until 1956 and 1958 respectively. The two former were an important innovation. The Constitutional Court, consisting of fifteen judges, was to decide on whether any law was consistent with the new constitution. Where it was not, the Constitutional Court could invalidate it and thus abolish a number of legalistic relics from the Fascist period. In future the careers of judges depended upon the Superior Council of Magistrates: they thus became, in theory at least, independent of the Minister of Mercy and Justice and therefore of governmental influence.

The new Italian constitution preserved compulsory military service. It also, in Article 44, interestingly stated that 'the law imposes . . . the transformation of the *latifondo* [giant landed property] and the formation of new productive units. The law assists small and medium property. Special measures in favour of mountainous districts shall be taken by law.'

It should be noted in conclusion that in the Transitory and Final Provisions at the end of the new constitution it was laid down that

(XIII) Reorganisation of the former Fascist Party, in any form whatsoever, is pohibited. . . .

Notwithstanding Article 48 [which said that 'All citizens, male or female, are entitled to vote . . .'], temporary limitations are established by law, for a period of not more than five years . . . on the suffrage and eligibility of the responsible heads of the Fascist régime.

All members of the House of Savoy were expelled and their property confiscated according to Provision XIII.

. Thus certain harsh reactions concluded the generally vague and utopian constitution.

In 1946 and during the first half of 1947, when the Constituent Assembly was being elected and during much of its work, conditions in Italy, both political and economic, were extraordinarily precarious. The partisans who had fought against the neo-Fascists embraced large numbers of communists, while it has been seen that many of the most influential partisans had been members of the Party of Action – believers, that is to say, in the liberal tenets of men like Carlo Rosselli and, another victim of the Fascists, the Piedmontese Piero Gobetti. The partisans had been backed by the industrial workers of northern Italy who already, under Gramsci's leadership over twenty years earlier, had demanded a share in management. From the positions of power which members of the Committees of National Liberation had seized they intended to introduce a new social order. Resistance on the part of the employers, some of whom like Falck had backed the partisans, was damned as Fascist.

Until May 1947 there were Socialist and Communist ministers in the Government and representatives of the Left had seized or been appointed to other key positions: for instance, in the spring of 1945 one of the leaders of the Party of Action, a southerner called Riccardo Lombardi, had made himself prefect of the province of Milan. He was an engineer by training and a man of infinite political ingenuity. He and others like him, in Turin and elsewhere, were determined that the 'second *Risorgimento*' in which they believed that the whole people had joined, should lead to real popular power, 'rule by the people'. By the spring of 1946 Lombardi had been replaced as prefect partly on the legitimate pretext of his fragile health. He survived as an active politician for many years, having become an irrepressible dissenter in the Socialist Party and the scourge of Nenni.

Power for the people was to have been realised, as Gramsci had planned, through the establishment of self-government in industry, active workers' councils. In the winter of 1945–6 and well into 1946 industry in Italy was almost at a standstill for lack of raw materials, power and transport, so that a share in management was something of an academic question. Moreover

Italy was at first in Allied occupation, and the British and American authorities were suspicious of the revolutionary aims of Italian labour and the possible influence of Russian Communism. This intensified Italian revolutionary feeling. Hordes of unhappy Italian prisoners-of-war returned from all the ends of the earth during this period. It is astonishing that such conditions did not provoke more explosion and violence than they did. The major factor in the stabilising of Italian society at this time was the relief supplied by the Allies on their own and by the United Nations through UNRRA. When De Gasperi visited America in January 1947 he was able to secure a loan of $100 million from the Export-Import Bank as well as help in shipping and coal. Only from May 1947 when De Gasperi, who, it has been seen, had led national coalitions since December 1945, succeeded in shedding his Socialist and Communist colleagues in favour of an almost entirely Christian Democrat Cabinet, was it possible to end the revolutionary phase. From this time onwards the remnants of CLN authority were destroyed, and for better or for worse the power of the Ministry of the Interior and the prefects it appointed was reasserted. The power of these prefects was modified, as it had been before Fascism, by the activities of freely elected communes and provincial authorities. It has been seen that regional authority above the prefects at first only emerged in special outlying areas.

Critics have claimed that the Allies with De Gasperi did not think out with sufficient care how politics and economics were to be co-ordinated in the new republic. Of course they did not, for they dealt with an emergency as best they could. If there was to be no Socialist revolution, then a liberal economy more or less on the British model must be accepted: De Gasperi felt that it would have to look leftwards as London itself was doing. Before Fascism an Italian Liberal leader like Giovanni Giolitti had bargained for the support of a largely Catholic electorate since the time when the papacy had sanctioned voting by Catholics. Giolitti had hoped to proceed from an oligarchical liberal economy to a democratic mixed one, but war and then Fascism had intervened. Now it was not so different. Though many employers recovered their positions, the emerging economy was a mixed one on account of the state holding concern, the *Istituto per la Ricostruzione Industriale* or IRI. This and some similar

organisations became an important though minor part of the new
Italian economy. The Government now, from May 1947 onwards,
was essentially Catholic in the sense of Christian Democrat, but
the electorate would be revealed as a paradoxical mixture. Those
parts of the country which had been Socialist in Giolitti's day
were emerging as Socialist or Communist, preserving their 'Red'
allegiance, while old Liberal areas voted Christian Democrat.
Although the Liberal Party after 1945 shrank to a relatively
small group of anticlerical conservatives, something of the old
Liberal tradition survived in the organisation of industry. Later
economic development was to push the whole of Italian society,
almost against its will, in a liberal direction, 'liberal' in the general
sense, which after all in his day Cavour had been, rather than
an Italian Liberal.

The Italian Socialist Party ever since its foundation in 1892
had constantly displayed its fissile or fissiparous nature. Indeed
it has been seen that one of its splits, in January 1921, had created
the Italian Communist Party which on the contrary proved itself
so stable. A split in January 1947 between the followers of Pietro
Nenni and those of Giuseppe Saragat was to have consequences
which may yet prove fatal to the Italian Republic; the results
may not be clearly visible until after 1970, although by then
Nenni himself was reconciled with Saragat who was President.
To Saragat political freedom was as important as social justice,
and he regarded with intense anxiety the Communist programme
of establishing a dictatorship of whatever the proletariat might
mean in reality. Nenni, on the other hand, had long been more
conventionally Marxist. It has been seen that he was convinced
that the Socialist Party must work in close alliance with the Communists. If political freedom had to be sacrificed to working-class
unity he was at this time willing to take the risk. He and the
Italian Communists accused Saragat of being bought by the
Americans just as Saragat regarded them as the tools of Moscow.
Thus the controversy merged into the Cold War background.

Saragat's attitude to the East–West conflict made it possible
for him often to join De Gasperi's governments: he had close
Socialist friends, like the writer Ignazio Silone, who disapproved
of this collaboration although they were equally anti-Communist.

Although these two main Socialist groups officially rejoined
in 1966, their reunion was hopelessly brittle and there were always

other Socialist factions, one within Nenni's party led by the irrepressible Riccardo Lombardi. The follies of the Socialists provide one explanation of the apparent strength and unity of the Italian Communist Party from this time on.

It should perhaps be noted that at its party congress in February 1946, Parri's Party of Action, descended as it was from Carlo Rosselli's movement, *Giustizia e Libertà*, had literally disintegrated. It cannot be said that this was due to personal ambition or rivalries; the Party of Action failed to establish clear lines of policy, although, perhaps because, it included many of the best, the most ingenious minds of Italy. Without the simplicity these people were unable to discover they could never attract sufficient popular support to make the party politically worth while. This was a sad thing for Italy. The *Azionisti* might have become a serious constitutional opposition accepting democratic rules and able to offer an alternative constitutional government. This was something which the republic was gravely to lack since, officially at least, the Communists desired no such role.

Saragat hoped to attract many *Azionisti* into the ranks of his Social Democrats, although on the whole they had drifted into the Republican Party which became more solid in consequence. It has been seen that in June 1946 the Socialists, still united, had polled a little better than the Communists; indeed in the so-called industrial triangle of Turin–Milan–Genoa they had gained 29·5 per cent of all the votes as against 21·9 per cent for the Communists. After January 1947 the latter were always the largest party in Italian political life, second only to the Christian Democrats. Naturally the Communists picked up extra votes from the split Socialists.

Immediately after the break between Nenni and Saragat the Italian peace treaty was signed. In fact, as Italian ambassador in Paris, Saragat had been closely concerned with it, and was probably affected by the behaviour there of the Russians. In spite of all Fascist propaganda Italian public opinion had never wished for intervention on Hitler's side in the War. Badoglio's declaration of war against the Germans in October 1943 and the splendid activity of the Italian partisans had encouraged the Italians to hope for greater generosity than they got from the Allies. The terms of peace with Italy were discussed by the Four-Power Council of Foreign Ministers in the autumn of 1945 and again between April and July 1946. A draft treaty was presented to the peace conference in Paris; it was approved by the end of the year and signed on 10 February 1947. The Italian armed forces were severely restricted. The Western Allies renounced all claims to reparations, but Russia demanded $100 million, and in addition a total of $260 million had to be paid to Yugoslavia, Greece, Ethiopia and Albania severally. Russia, the supposed inspiration of the Communist partisans, thus showed less understanding than the British and Americans. Into the bargain Moscow supported Yugoslav claims to a frontier west of Trieste (Istria was lost to Italy in any case) – claims which did temporary political damage to the Italian Communists. Until 1954 Trieste and its hinterland remained in Allied occupation. Meanwhile Italy's frontier on the Brenner was confirmed, leaving her with a German-speaking population of well over 200,000 in the Alto Adige. Ethiopia had recovered its independence, and Italy had also to renounce her pre-Fascist colonies of Libya, Eritrea, Italian Somaliland and the Dodecanese which included Rhodes. Later Somalia was placed under Italian trusteeship on behalf of the United Nations until it became independent in 1960. Territorial claims made by France were rejected except for some minute concessions on the frontier of the Val d'Aosta.

Although the peace treaty was a saddening experience for the Italians they accepted it with relatively little bitterness. They had for years lacked military ambitions in spite of all Mussolini's efforts, and the loss of their empire turned out to be advantageous, for they were spared the humiliations that decolonisation brought to Britain and France, and they saved the heavy expense in which their colonies had always involved them. In Somalia they established good relations with the Africans, and even in Libya, which became independent in 1951, the Italians were not at first unpopular. Here, too, the Fascist period seemed to disappear with strange completeness; the reappearance in a year or so of a neo-Fascist Party in Italy was not particularly impressive.

2 The Era of De Gasperi

By the time that the new constitution came into force at the beginning of 1948 a considerable stabilisation had taken place in Italy. Drastic credit restrictions imposed in August 1947 had forestalled what looked like being the collapse of the lira. When the restrictions were relaxed, imports rose and prices gradually fell. From the beginning of 1948 the value of Marshall Aid, originally offered in June 1947, was felt.

The first elections for a new Chamber and Senate, held according to the new constitution, took place on 18 April 1948. Four weeks earlier, on 20 March, the three Western Allies, the United States, Britain, and France, published a statement in favour of the return to Italy of the whole free territory of Trieste which had been created round that city between Istria, ceded to Yugoslavia according to the peace treaty, and Italy. The Left, both the followers of Nenni and those of Togliatti, denounced this statement as an unblushing attempt to exert political pressure upon Italy. In fact it is not certain that the election results were greatly affected by it – the Communist seizure of power in Czechoslovakia may have counted for more. The outstanding electoral result was certainly the success of De Gasperi's Christian Democrat Party which gained 305 seats out of 574, that is an absolute majority.

The Christian Democrats descended, essentially, from Don Sturzo's party of the *Popolari* which had been disowned by Pope Pius XI in the twenties, for he preferred Mussolini. Pius XII, although not more enlightened, accepted the change in circumstances which caused strong democratically Catholic formations to emerge after the Second World War in France and Germany as well as in Italy. *La Democrazia Cristiana*, it will be found, contained many differing political views; although closely related to the Catholic Church, few of its members accepted theocratic notions such as inspired Catholic Action. The latter was a highly conservative and authoritarian organisation of lay

Catholics; it was very close to the person of Pius XII through
its chief, Luigi Gedda. Interestingly enough there was a theocratic
tendency outside the circle of the Pope and the Curia among
a few tremendously idealistic Catholic leftists who took up an
evangelistic or perhaps Franciscan attitude. The true Catholic,
they believed, should follow St Francis in abandoning all worldly
goods, and the Church should rule society in this sense. The
remarkable Sicilian, Giorgio La Pira, who afterwards became
Mayor of Florence, was a Catholic of this kind.

On the other hand the parties on the Left lost very little in
April 1948 in spite of Saragat's defection from Nenni; they total-
led 216 seats instead of 219, though 33 of these were held by
Saragat's followers and in East–West terms were lost to the
East. In the Constituent Assembly there had been 115 Socialists
and only 104 Communists, but from this time onwards the Communist Party was larger than the sum of the Socialist parties,
undivided and very efficiently organised. The Christian Democrat gains had come, not from the Left, but from the Liberals,
the *Qualunquisti* and possibly even from the Republicans. After
this the *Qualunquisti* melted away; it was interesting that this
phenomenon lacked any staying power except in Fascist terms.
Already the people who had supported it in 1946 were veering
towards the now visible nucleus of Fascist reorganisation.

According to the new constitution the Chambers had jointly
to elect the President of the Republic. Provisionally since June
1946 the presidential duties had been exercised by an eminent
Neapolitan lawyer called Enrico De Nicola. In May 1948, however, Professor Luigi Einaudi was elected President. Einaudi, a
man of seventy-four, was an economist of great, indeed international, distinction. He had been made a senator in 1919 and
continued to attend the Senate in the Fascist period although he
had never concealed his dislike of Fascism.[1] He was a Liberal,
but far from being as inflexibly conservative as many Italian
Liberals. Indeed he always listened to the sometimes preposterous
comments of his old friend, Ernesto Rossi, a splendid anti-Fascist
enfant terrible whom no Fascist authority had been able
to tame throughout the Fascist period. Ernesto Rossi stood
much further to the Left than Einaudi, quite close to Parri.
It was a typically Italian situation that the conscientious

[1] For many years he was the correspondent of *The Economist* in Italy.

President paid such attention to Ernesto Rossi whose courage and wit appealed to him. Einaudi was a man of the greatest integrity and characteristically Piedmontese in his meticulousness, punctual, industrious, simple in his way of life. He considered that the President should be a neutral umpire but a very hard-working one who read all papers and dispatches in and out. Einaudi was a typical northerner.

The two other most prominent men in Italian life in 1948 were also from the north. Alcide De Gasperi had been born an Austrian subject in Trent, and Italians sometimes laughed at him as an Austrian. In his youth he had been prominent in Don Sturzo's party of the *Popolari* and in consequence he was *mal vu* by Mussolini. Work in the Vatican Library had protected him through the Fascist period. He lacked personal charm, but was a man of great sincerity, of moderation and persistence. People on the Left expressed disgust when he succeeded the picturesque and romantic partisan hero, Ferruccio Parri, as Prime Minister in December 1945; although they abused him as a clerical, this he never was. As far as possible he avoided purely Christian Democrat cabinets and any subservience to Pius XII or the Curia. Indeed he tried always to include representatives of the lay parties among his ministers: it speaks volumes for De Gasperi that he and Count Carlo Sforza became the best of colleagues. It is worth remarking that De Gasperi would have liked Sforza to be President rather than Einaudi.

Eforza, a Ligurian, was an extremely cultivated man, a delightful companion in spite of his notorious vanity which was childishly innocent. In his youth he had briefly been Foreign Minister under Giolitti from 1920 to 1921, when he did his best to conciliate the Yugoslavs in spite of the fanatical hostility of many Italians towards the Croats; this went back to the days of Croat soldiers in Austrian uniforms in Milan in 1848. Sforza said now that the Habsburg Empire was gone Italy should try to provide leadership for the new Slav states. He went into exile very soon after Mussolini was made Prime Minister, and lived mostly in the United States. The subservience of King Victor Emmanuel to the Fascist dictator disgusted Sforza who expressed his Republican views in no uncertain terms after 1943. This led to his famous clash with Churchill and delayed his return to power. It has been seen that he was elected President of the *Consulta*. Later for a short time he was in charge of the *Epurazione* or punishment of prominent Fascists, a fairly brief episode.

Early in 1947 when Nenni, on account of Saragat's challenge to him, resigned as Foreign Minister, De Gasperi appointed Sforza in his place, Sforza, an old anticlerical, the most prominent member of the Republican Party and with that pro-Slav record. He and De Gasperi worked together for four years, sharing enthusiasm for the 'European' idea; there was in any case strong feeling in favour of Europe among Italian politicians provided they were neither Marxist nor stood too far to the Right. De Gasperi and Sforza brought Italy into NATO from the beginning in 1949. Of course they were roundly abused by the followers of Nenni and Togliatti as the dupes of the Americans. Italy was also active in the Council of Europe – later Sforza's son was one of her representatives in Strasbourg – and she supported Robert Schuman's efforts from 1950 onwards which resulted in the creation of the Coal and Steel Community and then the Europe of the Six.

The problem of southern Italy had been shelved since 1870: dominated by three northerners, the Italian Government now tackled it. Italy south of Rome, historically shut away by the Papal States, appeared to be a hopelessly underdeveloped area, moving backward rather than forward, stationary at the best. On the whole soil and climate were bad; soil erosion had been encouraged by the cutting down of trees first to build the railways, such as they were, and on a far smaller scale by the Allied armies for their own immediate needs. Irrigation was a farce. The rivers were destructive torrents in the winter but dried away to nothing in the summer. Much of the land consisted of the *latifondi*, huge, neglected, extensively farmed estates belonging to relatively few and mostly absentee landlords. The peasants mostly worked as day-labourers on these estates, but there was usually only about a hundred days' work for them in the year. Until the Americans arrived with DDT, the prevalence of malarial conditions had caused them to live in squalid hill-towns, a long way from their work. It seemed as if no one had ever thought of trying to improve social conditions in southern Italy. When occasionally someone had tried to build a factory to provide work they found themselves creating a slum because too many people rushed in for the jobs. After the war there were often disturbances in the south;

egged on by Communists, peasants often squatted on the *latifondi*. It has been seen that Article 44 of the constitution, for so ambiguous a document, had condemned the *latifondi* fairly clearly. When De Gasperi went to Calabria in November 1949 it was said to be the first time an Italian Prime Minister had ever visited the south, and great hopes were raised. Obviously an attempt must be made to distribute the land more justly and to modernise farming methods: improvements in housing and the systematic creation of jobs in industry were desperately needed. In 1950 the Italian Government launched two schemes with the help of the American money which it was still receiving. The first was a programme for agrarian reform according to which the least well-farmed big estates were to be divided up, against compensation, among peasant settlers who were to be encouraged to farm more intensively. According to the second scheme, financed by a specially created fund called the *Cassa per il Mezzogiorno*, the south was to be 'developed', irrigated, fertilised and provided with new houses in new villages. Technicians, from the north of course, went south to teach new methods and explain new devices. Real efforts were made and much money spent with some beneficial results. Since DDT had abolished malaria there was no longer any need for the peasants to crowd into a hill-town like Matera with its frightful cave-dwellings. Carlo Levi's book, *Cristo si è fermato a Eboli*, had made Matera notorious in the very year of Italy's liberation, for it described his own exile to the south imposed upon him earlier by the Fascists.

It is difficult to make a clear analysis of the results of the reforming measures of 1950. By law over eight million hectares of land in the whole country were to be redistributed over the next ten years, but up to 1960 only about 8 per cent had actually been expropriated and not much more had been effected in the following decade – vested interests and practical obstacles had proved too great. Reform was only aimed at certain districts in northern Italy, the Tuscan Maremma, parts of the Po valley, but it was aimed at the whole *Mezzogiorno*. The planners believed that they had to help as many individual peasants as possible. In consequence they established a great number of small farms, often too small to be viable. And of course political patronage came in. The important thing was, however, that changes were initiated and southern society shaken out of its paralysis. And then irrigation was improved and dams were built and tractors and other machinery introduced. Small farmers were settled on the Sila plateau in Calabria before the end of 1950 and in 1953 a new village, La Martella, was built near Matera to which the cave-dwellers could move. It was characteristic that some of them were loath to do so and used their new cottages only to house their tools.

It was typical of the problem of the south that the provision of new machinery came from northern Italy which therefore, some people claimed, reaped the major advantages of the reform. It would, however, be entirely misleading to leave the matter there. Many more peasant farmers were established on the Sila and elsewhere during 1951, and by October 1952 it was officially stated by the Minister of Agriculture, probably with truth, that 107,000 hectares had been distributed to 23,000 peasant families. In 1955 great improvements were effected on the Ionian coast, round Metàponto and Sibari, where until now deplorable conditions had prevailed.

De Gasperi, although he launched the first big piece of economic planning in the agrarian reform and the beginning of the southern venture, and although he saw conditions created which would make great industrial developments possible, was already plagued by the instability of his governments. It has been seen that he valued the co-operation of the small lay parties with that of the big *Democrazia Cristiana*. The Liberals had fairly close links with the big industrialists which gave them a certain importance; they tended to shy away from that state interference in the economy which every day became more inevitable in twentieth-century society. When they found themselves in the same Cabinet as Saragat and his Social Democrat friends the situation became difficult, the Liberals hovering on the brink of resignation; in 1950 they did in fact resign, although they came back later. As for the Republicans, Sforza died in 1951, leaving Pacciardi as their leader. The latter served in several of De Gasperi's governments but became increasingly authoritarian; he ended up years later as an admirer of de Gaulle. It was Ugo La Malfa, deriving from the Party of Action, who gradually became the most prominent figure in the Republican Party. He supplied much expert financial knowledge and was already beginning to think in terms of long-term planning, not only in agriculture:

he first became a minister in 1950. It must be repeated that the *Democrazia Cristiana* itself contained all kinds of opinions within the common Catholic background. Some of the groupings, or *correnti* as the Italians called them, were socialistic in their thinking. Indeed, although in the days of Pius XII the notion was frowned upon, there were probably always a few Catholics who sympathised with Communism as being truly Christian.

One of De Gasperi's most extraordinary colleagues was Amintore Fanfani, a Tuscan who had as a young man taught at the Catholic University of Milan. Gemelli, its rector, was an enthusiastic supporter, not only of Fascism but also of National Socialist racialism; Fanfani had adopted and expressed such opinions as his own in the thirties. During the war, however, he turned against Mussolini and Hitler. He emerged in post-war Italy as a Christian Democrat with socialist and pacifist leanings, becoming the founder of a leftist *corrente* called *Iniziativa Democratica*.

Probably no one has played a bigger or a stranger part in the history of the Italian Republic than Fanfani, and his role is not yet played out. He first became a minister in May 1947 when the Marxists left De Gasperi's Government; Fanfani was then appointed Minister of Labour and threw his energy into the building of houses for the people – his was the only successful popular housing effort made by a minister of the republic before 1970. In 1951 he was moved to the then key position of Minister of Agriculture just after the launching of the great reforms. This put him at the head of the new personnel required to carry out the reforms. One of his Secretaries of State was Mariano Rumor, the future Prime Minister. The careers of Moro and Colombo originated similarly in the *Enti*, or groups of people engaged to work for the *Cassa per il Mezzogiorno*. Communist propaganda has exploited Italian credulity to suggest that the new jobs were based on corruption. For Italians are quick to think the worst of other Italians. Belonging to a society weighed down with poverty which is susceptible to bribery, Italians are too quick to suspect corruption when it has not occurred. Fanfani had a clever, tortuous mind and arrogant high-handed habits: he liked to build up power politically, but there is no evidence of corruption in his world, though possibly of pressures which could become intimidating in various ways. It was easier to whisper round suspicions

about Fanfani and reforms in the south because he had had a
southern, a Calabrian, mother. Later De Gasperi moved Fanfani
to the Ministry of the Interior, another key position where his
illiberal attitude was qualified by his sympathy with leftist reform.
Since the end of the war he had been closely associated not only
with La Pira, but also with a strictly Catholic leftist called Giuseppe
Dossetti, who believed in the sanctity of poverty and asceticism.
Just after the war in fact Fanfani had shared humble lodgings
with these other two in a house named after its sign of 'The
Piglet'.

Fanfani's character did not facilitate coalition government
which created a particularly intricate game in a society in which
political ideas were felt to be more important than mere political
action. The electoral system, a modified form of proportional
representation of the political parties, meant that so long as the
Communists and the major part of the Socialist Party were in
opposition, De Gasperi's governments, after the first electoral
period, could only retain a working majority in the Chamber with
the consent of the smaller parties or most of them. This involved
hair-splitting arguments, endless bargaining and much wasted
time.

Local elections in 1951 and 1952 alarmed De Gasperi as to
Italy's political future: he noted, not without bitterness, that the
extremes of Right and Left had gained support, particularly in
southern Italy. Even before 1948, in spite of the ban in the constitution, a frankly Fascist party calling itself the *Movimento
Sociale Italiano* (MSI) had been organised, holding its first
national congress in 1952. In addition a so-called Monarchist
leader, a Neapolitan shipowner called Lauro chiefly delighting
in his own publicity, had revived the Monarchist Party, which
also made gains in the early fifties. De Gasperi decided to search
for a path to greater political stability before the general election
due in 1953. He therefore introduced and carried a modification
of the electoral law so that any political list which was supported
by 50 per cent or more of the electors would be given a bonus of
seats (380 out of 590) in the Chamber. There was an immediate
outcry against this *legge truffa*, the cheating law. Not only did
it recall a more unblushing measure introduced by the Fascist
minister Acerbo, in 1923, not only did it seem to try to give the
Christian Democrats permanent power; it really shocked the

Italian sense of fair play in politics – the mathematical principle was more important to the Italians than that the machinery should work. In fact De Gasperi had overplayed his hand. The Christian Democrats just failed to gain 50 per cent of the votes for the Chamber in 1953, while the right-wing parties gained again in the south and the left-wing parties held firm. Soon after the election De Gasperi retired. He died in 1954: in the same year the *legge truffa* was repealed.

It seemed a tragic end to all his labours; southern Italy seemed to have spurned the great reforms he had initiated there, and all around him there was ingratitude. And yet the foundations were laid for great economic progress. It is difficult to believe that more extreme action could have achieved more for Italy in the first five years of the new constitution. A Marxist revolution would have been vetoed by the Vatican and the Western Allies: it would in any case have cost much more. With the Marxists in opposition De Gasperi had probably achieved, with his slender resources, as much as was then possible in the way of both social reform and the preservation of personal liberty; he had brought Italy back into the comity of Western Europe – already by 1952 the disarmament clauses of the Italian peace treaty had become a dead letter, for Italy was needed by NATO.

De Gasperi had in fact more or less restored what was by the mid-fifties the old-fashioned structure of the liberal state which Mussolini had done his best to destroy. Yet the liberal state must now be worked by the leaders of mass parties which did not believe in liberalism. The beliefs of both the Christian Democrats and the Marxist parties were flavoured with religious authoritarianism, the former that of the international, still conservative Catholic Church, the latter that of international Communism which by now ruled a large part of the world and in its Russian form was becoming conservative too. The functioning of parliamentary democracy in Italy after the middle of the twentieth century was warped by this anomaly. Ordinary life and economic progress in Western Europe pulled both Christian Democrats and Communists towards a common-sense type of political co-operation which in practice was liberal, but their ideologies pulled in the other direction. Since Italians, it has been seen, are on the whole more interested in political theory than in political practice, the Italian Parliament spent large amounts of time in formulating conflicting political ideas. As it became clear that the Communists would be in opposition for an indefinitely long period of time they settled down to professional opposition. Since the Christian Democrats allowed themselves to split into *correnti*, urgent new legislation was constantly delayed by faction, faction in practical alliance with Communist opposition. The more legislation was delayed, the more difficult it became to formulate and implement because great economic change after 1955 transformed the circumstances in regard to which legislation was required. The division of the Socialists only made matters worse. The other parties were all too small to affect the situation.

This deadlock was soon to be shaken by rapid economic change.

3 The Economic Miracle

According to the Italian peace treaty of February 1947 the region around the city of Trieste had been created a free territory of Trieste to be ruled by a governor chosen by the Security Council of the United Nations. However, the Council was never able to agree upon a governor: hence the Anglo-American occupation of the northern Zone A which included the city, and the Yugoslav occupation of the southern Zone B, dragged on. In effect Zone A was half assimilated into Italy and Zone B more than half assimilated into Yugoslavia. By the end of De Gasperi's time both the Western Allies and the Italians were tired of the tension in the so-called free territory. De Gasperi's successor, Giuseppe Pella, laid claim to the whole area, and at one moment mobilised Italian troops – only to arouse angry protests from Belgrade.

In October 1953 the Allies brought matters to a head by announcing that they intended to hand over Zone A to Italy, but Tito vetoed the Allied decision. Behind the scenes, moves, probably encouraged by President Einaudi, began to take place towards some kind of compromise, in spite, perhaps also because, of Pella's aggressive attitude. A long-drawn-out scandal about the death of a girl called Wilma Montesi, in which the son of the Foreign Minister, Attilio Piccioni, was (probably unjustly) incriminated, had one good result. It obliged Piccioni to resign from the Foreign Office. He was succeeded by Gaetano Martino who, unlike Piccioni, supported the President on the issue of Trieste. After months of negotiation in London between Italians and Yugoslavs under Anglo-American sponsorship, the most obvious terms were agreed upon. The demarcation line between Zone A and Zone B, with a few tiny adjustments, was accepted in October 1954 as the frontier between Italy and Yugoslavia. The agreement, described as a 'memorandum of understanding', was given as little publicity as possible. In the light of the past seventeen years it may be considered to have been the only happily settled post-war dispute. For relations between Italy and Yugoslavia have flowered richly. Nationalistic indignation, far from building up into any kind of pressure-group as was liable to happen in Germany, melted away in Italy, and the Fascist attitude of contempt for the Yugoslavs was forgotten. Very quickly exchanges of all kinds took place, and Italian, a language virtually banned in Belgrade between the wars, became popular there. Italian Communists and Socialists were interested in Tito's Yugoslavia and many Yugoslavs were glad enough to visit Italy, and even take jobs there: the Italy–Yugoslav frontier became one where no formalities were required. Italy was able to supply much that was needed for Yugoslav development. And the flourishing commercial relationship thus established made its contribution to Italy's new prosperity in the second half of the fifties.

Italy's great economic weakness until the middle of the twentieth century had been her lack of coal, iron, gas and oil: the development of water-power between the wars had not compensated for these shortages. In 1949, however, oil was struck near Cortemaggiore in the Po valley in Emilia, and natural gas was discovered at Lodi and near Ferrara. (Hitherto Italy had had to obtain her gas from imported coal.) These discoveries were made under the auspices of the *Azienda Generale Italiana di Petroli* or AGIP which had been founded by the Fascist authorities in 1926. In 1945 a Christian Democrat engineer from the Marche who had been an active partisan leader was put at the head of AGIP with the implication that he would wind up this Fascist organisation in the interests of the freedom of industry. But this engineer, Enrico Mattei, had quite other notions. He became convinced that Italy held important oil and gas deposits and he intensified research in order to discover them. Further he claimed that 'the state must keep its hands on the sources of energy that the state itself has discovered, and use this energy in the exclusive interests of the country'. In 1949 private interests pressed very hard against Mattei and AGIP, but he held his ground and was helped in 1950 by the resignation of the Liberals from the Government. In 1950 twice as much natural gas was produced as in 1949 – it was gas rather than oil which AGIP was able to provide. Mattei used its profits to subscribe to Christian Democrat funds; for this, too, he was of course furiously attacked.

But he had his way. Oil development went ahead also, most of all in Sicily where it was discovered in 1953,[1] but AGIP was able to make gas into Italy's vital source of energy. In February 1953, towards the end of De Gasperi's time, a law created the *Ente Nationale degli Idrocarburi* or ENI, a public corporation which was made responsible, in the national interest, for all activities concerning hydrocarbons in the Po valley.

As the chief of ENI, Mattei extended his activities and his power. ENI became in fact a huge holding company operating largely through other mixed holding companies. Among these AGIP was still to be found, soon to be joined by AGIP *Nucleare*, responsible for work on nuclear energy. ENI, in addition to its propagation of methane and other natural gases through its off-shoot SNAM, mined, drilled for oil, refined and distributed it, and conducted research; it also sold hydro-chemicals, textiles and heavy machines. In addition Mattei built and controlled the new industrial port of Ravenna near the ancient city. Further he con-trived that ENI should dominate a corporation engaged in motel construction, should own a publishing house in Milan and should bring out a new daily newspaper there; the latter, called *Il Giorno*, adopted a less traditional format than the *Corriere della Sera* of Milan or *La Stampa* of Turin, and supported Fanfani and the left wing of the Christian Democrats.

At the same time Mattei used his position and his power to gain great influence abroad. For in 1957 he entered into agree-ments with the governments of Egypt and Iran to survey for them and work their oil. And he created what was regarded as a pro-gressive 'anti-monopolistic' precedent by offering, not the 50:50 arrangement the international oil companies had traditionally offered, but one whereby ENI would do the job keeping only 25 per cent of the profits. Mattei thus established a special Italian relationship with a large part of the Arab world including Morocco, Tunisia and Libya. In 1959 he made a controversial agreement with Soviet Russia to import Russian crude oil as against Italian oil equipment of all kinds. Thus Mattei pursued a foreign policy of his own with much success; indeed he might be said to be pursuing Fanfani's policy after Fanfani had lost grip. Mattei, too, was exceedingly high-handed in his behaviour and was often described as the new Italian dictator: it is not

[1] See below, p. 107.

surprising that his support of the high-handed Fanfani through *Il Giorno* and ENI funds used directly, alarmed people of genuine liberalism. He of course said, as many technicians did, that one could get nothing done unless one rode rough-shod over the Italian bureaucracy. When Mattei was killed in an air accident in October 1962 some people breathed more freely. At first his successors seemed to lack his personality and drive – he was blamed for this too. He had in fact begun to be more conciliatory towards the international oil companies, and his successors followed his lead since Italy's need for oil was mounting, and inter-company disputes could but prove damaging.

It was in 1956 that it became evident that the Italian economy was beginning to boom; this was the year when the Italians got the commission to build the Kariba dam in Rhodesia. There is no doubt that ENI's contribution to Italy's development was of fundamental importance. Its own expansion was extraordinary at home and abroad; it included a major construction of pipelines for the transference of its wares. Already it impinged upon the great chemical industry of Italy grouped round the huge combine of Montecatini. The latter built a big new plant and a new port at Brindisi, an example of the contribution of southern development to northern expansion. Here at Brindisi oil could be brought in cheaply, contributing to a spectacular increase in the production of plastics.

The index of growth of the chemical industry went up from 100 in 1953 to 273 in 1961.[1] Closely related products, such as cellulose, paper and rubber, also increased on an impressive scale, cellulose and artificial fibres expanding three and a half times between 1953 and 1961, and paper and rubber nearly twice.

Equally remarkable was the development of the car industry. Here 89 per cent of Italian cars were produced by Fiat and exported all over the world. Fiat was now directed, not by the Agnelli family which had founded it and returned to its direction later, but by a remarkable figure called Vittorio Valletta who seemed to have Mattei's vigour without his ruthlessness. Under his leadership Fiat developed its production of lorries, tractors and diesel engines. Enrico Piaggio had seen to it that motor-scooters, immortalised as Vespas or 'wasps', made their contribu-

[1] See S. Clough, *The Economic History of Modern Italy* (1964) pp. 317 ff,

tion: the production of these had increased phenomenally immediately after the war and showed an index growth of 166·5 in 1961 compared with 100 in 1953.

The development of typewriters and other office equipment and sewing-machines was even more sensational than that of cars and motor-cycles, the best-known name here being that of Olivetti, a greatly respected Jewish family. The original Olivetti factory at Ivrea in Piedmont was well known for its paternalistic care for its workers and it was proportionately disliked by the Communists. Adriano Olivetti, the head of the family after the war, was full of plans for extending social welfare. After his death in 1961 the firm ran into difficulties in attempting to expand on the world market.

In the boom period, in spite of the lack of iron supplies at home, Italy greatly increased its production of steel and developed an export trade in certain specialised products such as pipes for pipelines. Steel production, according to the plan of the steel-king, Oscar Sinigaglia, was centred mainly on Cornigliano near Genoa. To help the south, the new *Finsider* steel plants were created, mainly thanks to IRI,[1] at Taranto, and, earlier, at Bagnoli near Naples. It should be recorded that shipbuilding went rapidly ahead though less steadily. Multiple stores like the *Rinascente* flourished in this boom period.

All along the basic production, by various means, of electricity was expanding at a great pace: before nationalisation this industry was dominated by the huge Edison concern, one of whose claims to fame was that it given Parri employment in Fascist days. Building developed on a grand though ill-regulated scale: it included the construction of a great number of hotels in conjunction with the extension of tourism. Here, too, efforts were made to help the south, for the woollen manufacturer, Marzotto, built his *Jolly* hotels mostly there. So densely did buildings of all kinds for the use of tourists develop in some areas that one began to ask oneself whether this scale of *sviluppo* (development) would not prove self-defeating; it continued throughout the sixties. Around the north of the Adriatic the coast became unrecognisable and the visitor who wanted to admire, say, the cathedral at Grado was forced through whole thickets of *pensioni* and hotels. Many new roads belonged to this part of the story, including the great

[1] See below, p. 30.

Autostrada del Sole from Milan to Florence and further on to the south via Arezzo; people said it went through Arezzo to please Fanfani who was born there.

Italy started her boom period with at least two million unemployed as well as much underemployment. However much the unions agitated, this original fact meant low wages, and therefore low labour costs which facilitated competition abroad. Then at the end of the fifties the European Economic Community began to bring advantages. Among others it made possible better conditions for the large number of Italian workers abroad; many of these worked in Western Germany but also in Switzerland which was not a Community country but more or less accepted Community standards. The Italian workers abroad sent home remittances sufficient to affect Italy's financial situation very favourably, although generally rather less than tourism did. In the fifties the Italian public saved rather than bought: this was a factor in the provision of investment. Total investments in the Italian economy in 1959 were 10 per cent higher than in 1958, and until the end of 1962 there was a progressive increase each year.

In later years much was said about the lack of planning in this period. There was, however, more planning than met the eye of the outsider owing to the existence of the *Istituto per la Ricostruzione Industriale* or IRI, a state holding company which had been founded in 1933 to save slump-threatened banks and industries, and which, like AGIP but unlike the Fascist corporations, had survived. In the fifties it owned a considerable portion of Italian shipbuilding and of Italy's mechanical industries; it also owned the whole telephone system and *Alitalia*, the main Italian airline. After the end of the war men in whom the governments had confidence (rather like Mattei) were given leading posts in IRI, occasionally some old anti-Fascist stalwart. The Government's nominees to such positions were able to guide investment a little. At first they were said to be in the pockets of *Confindustria*, the industrialists' organisation, as IRI had been under Mussolini. But this, if ever true, soon ceased to be so, and the IRI people were able to introduce a certain amount of government direction. These men lacked the absolute control Mattei enjoyed over ENI, nor were they as ruthless; private enterprise continued to compete with nearly all the concerns they controlled and sometimes confounded their plans. Even so there were overlappings; Monte-catini, certainly at one point in the first boom period, was not quite independent of IRI. At the end of 1956 the Segni Government placed a Ministry for State Holdings or Participation over IRI, ENI and the *Cassa per il Mezzogiorno*, and from 1958 onwards the Ministry had the last word over the IRI concerns. Over ENI it did not prevail while Mattei was alive.

The boom years in Italy, the years of the economic miracle, transformed the country. It has been seen that an enormous amount of building had taken place, too much of luxury flats and not nearly enough housing for working people – here the plan-lessness was flagrant. It has been seen that the great new *autostrade* were planned and begun. Unemployment ceased to be a major problem for the first time for half a century, and simpler people gradually begun to buy things that would hitherto have seemed remote luxuries to them. The whole feeling of Italian life changed. Not only was the sharp edge of poverty blunted, but the structure of Italian society, still relatively static even in the north, suddenly melted and became fluid. Instead of the majority of Italians working on the land, by 1960 the majority worked in industry. Italy had been essentially a land of small towns with famous names, capitals of provinces, with many small shopkeepers closely related to the country people: a huge city like Milan was still exceptional in the first half of the twentieth century. The Fiat factories had been kept outside Turin which in consequence did not make the impression of a giant town. The double capital of Rome had been swollen by Fascist centralisation: post-war bureaucracy and political life continued its extension. But by 1960 Milan, Turin, Rome and Genoa had grown much bigger, and the population was shifting as it never had before, mostly in a nor-therly direction.

The multiplication of industrial jobs mainly in northern Italy had differing effects in different parts of Italy. In southern Italy the *Cassa per il Mezzogiorno* seemed to work very slowly and many peasants who had applied for new plots on expropriated land were disappointed. Politicians from both Left and Right told them that they were being cheated by the Christian Democrats. The young men, if adventurous, had gone off for temporary jobs to Switzerland ever since the war had stopped; later they had gone to Western Germany too. Now in the second half of the fifties

they heard of jobs in industry in northern Italy where there would
be no foreign language to learn or foreign currency to understand.
So there was an exodus of southerners to the northern cities just
when it was planned to open new factories in the south – but of
course no one in the south believed in such plans.

One of the difficulties between north and south Italy ever since
Italy was united had been that southerners, because they were
poor, had taken ill-paid small official posts in the north or in
Rome, posts which northerners despised or did not need. Milan,
Turin and after 1918 even Bolzano always seemed full of policemen or postmen or tax collectors from the south. But these had
been mostly white-collar workers who had to be able to read and
write and thus did not come from the really poor homes in the
south. Now illiterate young southern peasants applied for jobs in
industry in the north; they applied in hordes. A city like Milan
with its good traditions had recovered from the war by the middle
fifties; it had got its people housed and its children to school, and
illiteracy was almost forgotten until this influx. It was painful for
the Milanese to see their standards destroyed, at least as painful
as it may have seemed for people in Britain to feel that an alien
coloured population was threatening their 'way of life'. The social
confusion was tremendous. It did not pacify the southerners to see
how much richer the north already was and how much richer it
was becoming.

In spite of the tremendous changes going on, Italian industry
did not become as concentrated as that of France. Many smaller
concerns in fact expanded during the boom. This was particularly
true in central Italy, the Red Belt, which was in consequence
transformed and yet preserved both at the same time. In Emilia
and Tuscany and, to a smaller extent, in the northern Marche
and Umbria, the land, except on the Apennines, was fertile, and
the peasants had been comfortably off. Many of them had shared
their crops with their landlords, but although leftist politicians
agitated against this *mezzadria* system it had often created satis-
factory relationships. These were the Chianti-growing peasants.
With the boom years the younger generation mostly left the land
to go into small local industries. Crop-sharing was under dispute
and therefore sounded unattractive to the young, though it had
provided their parents with working capital. The peasants who
remained on the land became more commercialised, selling their

wine and olive oil more systematically – in fact they often became part of the working of a new concern: almost certainly their land was bought up by one when they died. Poor land at any altitude, most often in Umbria, was simply deserted; probably the peasants who had worked it went into industry without feeling that they had lost their roots if the industry was local. A beautiful old Tuscan hill-town like Cortona, which in spite of its Communist mayors had failed to found industry on any scale, seemed to die on its feet. It was too remote, not sufficiently proverbial to attract more than a trickle of tourists. Less and less peasants came from outside to sell their vegetables in the market on Saturdays. Thus the Red Belt on the whole changed less than the north or the south; no rush of immigrants, no sudden exodus. It will be considered later how this affected political opinion.

4 Political Transition

Meanwhile Einaudi's term of office as President had expired in 1955 and the Senate and Chamber proceeded to elect his successor. The official candidate of the Christian democrat party was Cesare Merzagora, the President of the Senate. A series of intrigues succeeded, however, in bringing about the election of Giovanni Gronchi. Gronchi had represented his native Pisa in the Chamber before the Fascist era: he had been a member of the *Popolari*, the democratic Catholic party of Don Sturzo and De Gasperi, and he was prominent in the world of Catholic labour until the Fascist authorities put an end to such activities. The election of Gronchi was welcomed by many liberal-minded people as offering a chance of greater progress in Italian politics, for Gronchi was known to agree with Saragat in advocating some kind of 'opening to the Left'. This phrase, which was to become exasperatingly familiar in the following years, indicated the idea of drawing the Christian Democrats, or at least those with more flexible notions, into co-operation not only with the tiny Republican Party and that of Saragat, but also with the general body of Socialists, the *Partito Socialista Italiano* or PSI. It was a formula devised to achieve new social, even socialist, reforms; by doing so it was hoped that their thunder would be stolen from the Communists and that so at last their strength might be broken. For some time Nenni had shown himself to be not impervious to a formulation of this kind. The reforms envisaged would be based upon the idea of national planning, first publicly, indeed officially, formulated earlier in this same year by Ezio Vanoni, Minister of the Budget in the Cabinet of the conservative Christian Democrat, Mario Scelba. The Liberals in this Government, led by an able and upright politician called Giovanni Malagodi, objected to Vanoni's approach. Malagodi was in fact the ally of the industrialists who so bitterly criticised Enrico Mattei's activities. With the election of Gronchi, who was to prove more ambitious than Einaudi, indeed almost an imperious President, pressure from the Quirinal[1] would be exerted on the side of Vanoni and Mattei against the private industrialists. For a start Gronchi asked an experienced politician, Antonio Segni, to form a new Government in the place of Scelba: Segni remained in office as Prime Minister until May 1957.

Meanwhile with economic expansion beginning to surge ahead, bringing inevitable social dislocation, the news seeped through of Khrushchev's revelations at the Twentieth Soviet Party Congress in February 1956 and of the beginnings of de-Stalinisation in Russia. There followed the news of the 'Polish October' and of the rising in Hungary against the yoke of Soviet Russia. The Italian Left was for the time being profoundly shaken, and shaken apart. For Pietro Nenni the shock was startling. His alignment with the Communists seemed to him to have been based on false assumptions, and Saragat's policy seemed more justified. Already in August 1956 before the Hungarian rising he had had a meeting with Saragat at Pralognan in the south of France to discuss the possibilities of Socialist reunion, a meeting which for the time being was fruitless. But Nenni had begun to turn his back on the Communists, and this proved something of a landmark in Italian politics, though Nenni's Socialist Party followers were far from united in support of him.

For the Italian Communist Party the revelations of 1956, quite apart from Nenni's defection, spelt something like dismay. The Italian Communists, thanks in part to the most cherished founder of the party, Antonio Gramsci, had always preserved a more 'human face' than their colleagues in Russia. They had in fact never believed in the abominable performance of Stalin. Now they were obliged to do so and to write it off as a fearful deviation. Their habitual supporters in central Italy paid little attention, but many more critical voters turned away from them, and all their critics and enemies were encouraged: there was much sympathy in Italy on behalf of the Polish protesters and the Magyar rebels against Moscow.

Segni was a handsome Sardinian landowner with a reputation for being enlightened. He had been Minister of Agriculture at the time when De Gasperi initiated the agrarian reforms and he had supervised the distribution of some of his own land to

[1] Where the President resided.

Sardinian peasants. Apart from upheavals in Eastern Europe and the embarrassment of the Suez affair to Segni's government, his period in office was relatively uneventful at home. There was much discussion of reforming the *mezzadria* system characteristic of land tenure in central Italy, but the discussions furnished an example of the growing tendency of the Italian Chamber to talk at length without resulting legislation. The leftist parties and the Christian Democrat Left wanted to abolish the traditional sharing of crops in order to make the small farmers of Tuscany independent, but the Christian Democrats of the Right and the Liberals fought to preserve the big owners' rights. By now the Government was hopelessly split: when Segni resigned it was the end of the *quadripartito* or four-party government which had – intermittently and impossibly – combined the Liberals with Saragat, the Right with the Left.

A caretaker one-party Christian Democrat administration under Adone Zoli took over in order to hold the governmental fort until the general election of May 1958. When the voting took place it was clear that events in Eastern Europe had had their effect. The parties of both Saragat and Nenni had gained, but the Communists lost only three seats. The Christian Democrats gained twelve which, in the circumstances, was not considered surprising; the most probable explanation was that the energetic Fanfani had been Secretary-General of the Christian Democrat Party since 1954 and had organised it as it had never been organised before. Ground on the Right that would hardly be won back was lost by the Monarchists and the MSI, or *Missini*, as voters for those Mussolinian initials were called.

Under pressure from Fanfani the left wing within the Christian Democrat Party had gained. Gronchi now appointed this stormy figure to be Prime Minister. Fanfani remained Secretary-General of his party; in addition he took on the Foreign Office where he took care to push forward supporters of his own. His links, financial and other, with Mattei were evident. Judged by the resentment aroused by these two dictatorial men in combination, it is indeed surprising that their Christian-Socialist programme was not damned well in advance of what were officially recognised as the Left-Centre governments of the sixties. Mattei's international operations, whereby he arranged to exploit the oil of Iran and Egypt on such generous terms, were encouraged by

Fanfani. It was, fortunately perhaps, Fanfani who was voted out of power in the Chamber at the beginning of 1959. President Gronchi, who approved of the Fanfani–Mattei combination and in his way – he was a lesser man – was similarly aggressive in spite of his supposedly neutral position as President, wished to keep Fanfani in office. This proved impossible thanks to the voting in the Chamber but also to Fanfani's temperamental behaviour, for he resigned all his offices including that of Secretary-General to the Christian Democrat Party. His wholesale resignation had historic consequences in that his successor in the party in March 1959 was Aldo Moro, a professor from the University of Bari of whom little had been heard before. Moro was calm and tenacious, a quite different character, prominent in the moderate Christian Democrat group of the *Dorotei* who were named after a convent where they met.

The next two or three years of Italian history were utterly confused and tumultuous. The economic boom was gaining force with all that that involved. Increasingly it was debated as to who should control this development, the old-type industrialists or representatives of the state – it was in 1958 that the Ministry for State Holdings became effective in Italy's economic life. The opening to the Left, of which more and more was heard, was blocked at least as much by the Roman Catholic Church as by private industry. This was one reason for the chronic division within the Christian Democrat Party and for the *immobilismo* of all Italian governments since De Gasperi's resignation. The Republican leader, Pacciardi, became so exasperated that he began to support the idea of introducing a Gaullist system into Italy.

For the Church as represented by Pope Pius XII was authoritarian and conservative, nearly as hostile to the liberal spirit as it had been under Pio Nono. Pius XII had always been profoundly anti-Marxist and in his mind any kind of modern socialism seems to have appeared sinful. Altogether his approach was rigid and narrow. In July 1949 he had excommunicated all Marxists, and through Catholic Action and other Catholic associations, as through the hierarchy and the priesthood, the Pope instructed all the voters he could reach to support Christian Democracy, really meaning its right wing. Moreover, in spite of Togliatti's reticence, bitter left-wing criticism of the Church was induced by the well-known facts that it owned property on a big scale and held big investments in industry. This was no doubt inevitable; it had, after all, the expense of the upkeep and preservation of some of the most grandiose buildings in the world. On the other hand some of its landed property was believed to contain some of the worst Roman slums.

It seems probable that the Vatican, certainly a few years later, controlled or had large interests in five banks, four investment companies, the steel corporation *Finsider*, three property companies, three building companies, two flour and spaghetti companies and others.[1] In 1968 the Italian Minister of Finance estimated that Vatican investment in Italy amounted to 100 milliard lire; those in the United States were less than half as big. Of course the *Osservatore Romano* would hear nothing of these assertions, but the *Osservatore* was, after all, an instrument of propaganda. Catholic pressure on education was strong and intolerant. Above all the Catholic ban on divorce created much personal misery and doomed many children to hopeless illegitimacy. What made matters worse was that it was generally known that it was possible for rich people to compound with the Church on this issue.

Feeling became exacerbated early in 1958, before the elections of that year, in anticlerical Tuscany, where the Bishop of Prato accused a husband and wife of living in sin because they had not been married in church, but only by civil law. A case for slander was brought against the Bishop which in the first instance he lost. Later the court concerned declared itself not competent, but in the meantime the Vatican had excommunicated all those responsible for the trial. On 3 May 1958, three weeks before the elections, a letter was published in the name of the Italian bishops' conference – it is not clear that this had met – urging the public to vote 'in conformity with the principles of the Catholic religion'. The indignation which this appeal generated was kept alive by many incidents. Fanfani, who as Secretary-General had reduced the influence of Catholic Action in the Christian Democrat Party, felt himself to be persecuted because of his leftist policy by ecclesiastical figures such as the highly traditional Cardinal Ottaviani. In the period leading up to the extraordinary crisis of 1960 many Italians felt that the Church was on the wrong side of things.

[1] Corrado Pallenberg, *Vatican Finances* (1971).

In April 1959, indeed, the Holy Office, just when the idea of the opening to the Left was gathering force, issued a decree condemning political collaboration with the Socialists as with the Communists. In January 1960 Cardinal Ottaviani preached a sermon in Santa Maria Maggiore in Rome harping on the same theme and evidently condemning the visit to Moscow planned by the President of the Republic. Ottaviani was of humble birth and simple habits, an old man going blind; about the society in which he lived he remained unblushingly conservative.

It has been seen that in spite of the ban in the constitution a Fascist party had been more or less reconstituted as the *Movimento Sociale Italiano*. It had held its first national congress in 1952 and had done quite well in the elections the next year, but it lost noticeably in 1958. In the Chamber the MSI deputies were more or less ostracised – it was understood that no government could accept their support. Zoli in early 1958 had done so briefly but not with impunity. Fanfani's fall in January 1959 created a real *impasse*. The political impetus towards the Left or Left of Centre was blocked by the opposition of the Church. After an attempt with Segni again as Prime Minister, Gronchi, under a wrong impression, appointed a former Christian Democrat Minister of the Interior, Ferdinando Tambroni, to succeed him. Tambroni turned out to be ambitious and unscrupulous. He thought he espied power for himself to the Right. When he accepted the support of the MSI, without which he had no majority, three of his Christian Democrat colleagues, Sullo, Pastore and Bo, promptly resigned.

The MSI now arranged to hold its party congress at Genoa at the beginning of July 1960. On 30 June, with remarkable spontaneity, people from every kind of anti-Fascist party demonstrated against the MSI. Genoa had a deeply anti-Fascist tradition, and the idea of the congress was unbearable. From Genoa disturbances spread naturally to Bologna and Reggio nell' Emilia where there was a general strike and ten demonstrators were killed. The whole atmosphere was profoundly serious. The Italians knew, as they had known in 1898,[1] that evil developments threatened but must be prevented. Thus at least twice since unification the Italians have unerringly distinguished between

[1] In 1898 there was a remarkable demonstration of solidarity against the extreme Right then in power.

the darkness and the light. Tambroni was quickly defeated and disappeared ignominiously.

The extraordinary thing was that Pius XII, who died towards the end of 1958, had been succeeded by a pope whose attitude was entirely different. John XXIII, formerly Cardinal Roncalli of Venice, had shown French sympathies and sympathy for the Italian Socialist Party. For the first year or so of his papacy he remained in the background and allowed, or rather did not prevent, the rightist excesses of Cardinal Ottaviani and his friends in the Curia who had shown sympathy for the neo-Fascists. But in April 1961 the new pope published the encyclical *Mater et Magistra*; he was preparing to summon a General Church Council, the first since 1870, for the following year. Meanwhile Mattei, through his paper *Il Giorno*, was urging the desirability of the opening to the Left. Another important paper now speaking in its favour was *La Stampa* of Turin, partly owned by Fiat; indeed the enlightened director of that concern, Valletta, used all his influence to help bring about a Centre-Left Government.

5 The Opening to the Left

After the Tambroni crisis, and with the other parallel anxiety created by the apparently renewed strength of the Italian Communist Party, co-operation between the Christian Democrat reformers and not merely the Social Democrats, but Nenni's Socialist Party, obviously came appreciably nearer to realisation. The slow progress towards this goal can be partially explained in terms of the personalities involved. It has been seen that Fanfani was excessively touchy while Nenni was a character made up of weaknesses caused by lack of intellectual clarity; he was tormented, partly because he was good-hearted, by the highly intelligent Riccardo Lombardi who refused to leave the PSI but yet from within it invariably condemned all Nenni's decisions. Nenni, with much difficulty, had decided to break with the PCI in spite of his entanglement with the Communists in the CGIL, in spite that is to say of his own pro-Communists or *Carristi* (as they were called) who opposed the breach at any cost. Indeed Luciano Lama, Secretary-General of the CGIL, at this point left the PSI and became a Communist. There remained for Nenni to decide whether he could accept Italy's membership of NATO which the Christian Democrats would not abandon, and whether he could accept the social policy of men like Fanfani and Moro. (It would probably not be exact to ignore a pardonable desire on Nenni's part not for ever to remain in opposition.) Did they sincerely favour planning? Did they really intend to nationalise electricity in spite of its flourishing condition, i.e. to do so on principle? Papal influence was now working strongly in favour of the opening to the Left which was first put into practice by the formation of a Socialist–Christian Democrat *giunta* or town council in Milan and Genoa and in one or two other cities in 1961; the region of Sicily also established a Centre-Left government in the same year.

Now the leftist Christian Democrats expressed a growing feeling that the economic boom had been half wasted because it had not been planned, though of course most of the entrepreneurs and the Liberals claimed that free enterprise was responsible for the expansion and should not be curbed, least of all at this time. It has been seen that since 1958 a Ministry for State Holdings had directed a good deal of economic activity, nominally even that of Mattei. At a Christian Democrat convention at San Pellegrino in September 1961 Professor Pasquale Saraceno, the leading party economist, brought forward a report supplying expert backing for the demand from the Christian Democrats on the Left for economic planning.[1] The remarkable economic expansion of Italy was being allowed to increase the economic discrepancy between north and south, since the north was growing richer faster, and money which was needed for investment was simply being spent on the acquisition of consumer goods. Therefore, Saraceno said, a planning office should be created. There should be fiscal reform such as to encourage development and to restrict private gain. The modernisation of agriculture should be pressed forward with the abolition of the *mezzadria* and through well-thought-out credits, and, finally, electricity should be nationalised because it comprised a national service that should, as Mattei had said of gas and oil, be at the disposal of the community: after all many other industries depended on it.

The Socialist reaction was an economic paper published early in 1962 making it clear at last that Nenni and his supporters were willing to collaborate with the left-wing Christian Democrats in working for change from within the Italian state without demanding its overthrow and a fresh start. The Socialists' economic paper echoed Saraceno's report, placing great emphasis on the need to nationalise electricity as part of planned economic growth, and also on the extension and modernisation of education. At the Christian Democrat Congress immediately afterwards Moro repeated all these aspirations. In February 1962 Fanfani formed his fourth government with these joint Christian Democrat–Socialist aims as his programme. It is important to make clear that Nenni and his party were not included in this government, but they undertook not to oppose it.

In fact Fanfani made a far-flung announcement. First of all, regional administration was to be introduced throughout Italy,

[1] See J. La Palombara, *Italy: The Politics of Planning* (1966), p. 64.

that is to say fifteen new regions were to be created. In the constitution the regions had been allotted wide powers, with hints of planning. Somehow many people with leftist sympathies thought of the regions as ideal areas for planning, and to the Socialists planning was almost inseparable from regional government; they had made the introduction of both the price of their support. Fanfani also announced far-reaching social reforms in the direction of housing, health and particularly education. Finally he declared that his government would within three months present a bill to Parliament to bring about a 'rational unification' of electric power in Italy.

After four months of heated debate Fanfani's bill for the nationalisation of electricity became law in November 1962 with very little amendment. All private ownership of any kind of electricity was expropriated. Compensation was to be paid to the companies concerned, not to their individual shareholders, over a period of ten years at an annual interest rate of 5·5 per cent. The management of the nationalised electricity was entrusted to an *Ente Nazionale di Energia Elettrica*, its name usually shortened to ENEL, a public corporation which was not subjected to the control of IRI or the Ministry for State Holdings but was to be regarded as subject only to the Ministry of Industry and Commerce. This, it seems, was intended as a mystical tribute to ideology; ENEL was to be placed above mere production or productivity. But it is impossible not to feel that ENEL should, together with ENI, have been subject to the Ministry for State Holdings, for it was necessary to co-ordinate the activities of ENEL with those of ENI as was later to occur.

The nationalisation of electricity was the most solid achievement of Fanfani's fourth government: in spite of so much emphasis on the regions to be, and a great deal more discussion of them, only a fifth one actually emerged, Friuli–Venezia Giulia with Trieste as its capital, towards the end of 1962. One critical and one major development had, however, been brought about. The legislation on electricity caused a flight of capital from Italy: the boom was beginning to die down and this hastened its end, though not, of course, the end of what it had achieved. The major change was the incorporation in government policy of planning as a principle, epitomised in the transformation of the Ministry of the Budget into the Ministry of the Budget *and*

'*Programmazione*', the latter word being preferred as less contentious than *Pianificazione*.

The new principle raised all the issues involved in putting it into practice. Already the thing that probably worked worst in Republican Italy was the civil service, for the reform of which one of Fanfani's colleagues, Giuseppe Medici, was at the time responsible. Planning was bound to tax the unfortunate bureaucrats much more heavily; even if new people were appointed to the new jobs, much more co-ordination was bound to become necessary. As it was, the various ministries tended to be isolated from one another. They employed too many ill-paid and ill-trained officials, often southerners who, by wallowing in red tape, took it out of other citizens that they felt inferior towards them. The very nationalisation of electricity had created a tangle of the authority of the Ministry of Industry and Commerce over ENEL with that of the Ministry of State Holdings over ENI and IRI and the *Cassa per il Mezzogiorno* with its fiefs, the southern *Enti*. Now, since planning had been made the business of the Minister of the Budget, he and his officials would expect to be, or to behave as, co-ordinators of all the rest.

The man appointed by Fanfani to this position in February 1962 was Ugo La Malfa, an exceedingly intelligent Sicilian who had belonged to the Party of Action in order to resist Mussolini and had recently succeeded Pacciardi as leader of the Republican Party. Like so many Italians, La Malfa had originally been in love with political theories. He had learned much from experience, however, and had become a mature Liberal in the English sense. Planning as conceived by him would begin and end with parliamentary consent. On 22 May 1962 La Malfa presented his *Nota Aggiuntiva*[1] on the whole subject to the Chamber by which it was approved. In it he showed how the unbalanced proportion of recent production in north and south Italy required a change of policy over to planning. 'The new and old problems of our economy now require more resolute action than in the past and above all a daring new approach, assured in fact by the choice of a policy of general planning.' The approval by the Italian Parliament of La Malfa's *Nota* was 'the first time in the history of the Republic that budgetary debates concerning other sectors were preceded by specific legislative decisions concerning economic objectives'.

[1] La Palombara, p. 93.

The next thing La Malfa did was in August 1962 to establish a *Comitato Nazionale per la Programmazione Economica* (or CNPE) under his own formal chairmanship as head of his ministry, but with Saraceno as vice-chairman and chief of the experts nominated to the committee by the minister. On the committee which got to work in September, there also sat representatives of capital and labour, that is to say of industrialists, landowners, bankers and trade unions of differing political complexions. It was interesting that, although most employers were hostile to planning, their representatives were not unwilling to co-operate on this committee. But here were the seeds of new administrative problems, particularly because, if Italy wished to remain a free country, the co-operation of people hostile to planning had to be induced, not enforced. All the Italian ministries, though overstaffed, were short of well-classified information which was essential for planning at any level; indeed civil servants often depended on industrialists and their staff for information just when the industrialists were mostly inimical to effective planning. If the regions were to be realised and to become the chief vehicles for planning, as the planners hoped, they were likely to lead to the creation of new officials, although the enthusiasts for the regions hoped that the regions would reduce the number of central government officials, and possibly eliminate the provincial authorities or at any rate the traditionally unpopular prefect of each province. In 1971 we have yet to see.

In view of the existing state of administration and the new problems created by the nationalisation of electricity alone, the industrialists of *Confindustria* had a good case against planning. Even if they themselves had helped to shake confidence, the newspapers they influenced like the *Corriere della Sera* and the *Nazione* of Florence could credibly protest that the boom was being killed by Fanfani's policy. Some of the people on this side of things thought that official planning should be restricted to the public sector of the economy – IRI, ENI, the *Cassa* and so on, and now ENEL. The Liberal Party held its party congress in April 1962 before Fanfani had introduced his bill for the nationalisation of electricity. It has been noted that the secretary-general of the PLI or Liberal Party was the able and upright Giovanni Malagodi, son of a distinguished politician who had been close to old Giolitti. He formulated the position of his party as: approval of planning in the public sector, and approval of the making public by the Government of the country's economic situation in general with 'emphasis on the prospects for development in the productive sector'. The Government must not, however, have any power of coercion outside the concerns over which it had control. This statement had a good deal of influence: after all the Government itself was not claiming the right to compel except where it nationalised. At the same time the Liberal Party came out unequivocally against the establishment of the regions throughout the country – the regions to which the advocates of planning attached so much importance.

Towards the end of 1962 the economic outlook was uncertain but certainly stormy: at the same time the stormy Prime Minister was becoming irritated by the obstacles he encountered. Most Christian Democrat politicians preferred temporary caution in view of the general election which was looming in sight. For the time being the *Apertura a Sinistra*, and with it Fanfani, went out of favour; Fanfani felt bitter also over the fact that earlier in the year Segni, whom he regarded as a rival, had been elected President of the Republic to succeed Gronchi who in his clumsy, over-eager way had contributed to the achievement of the *Apertura*; Segni was a much more prudent man, though not necessarily an illiberal one. Later, when he could not defend himself, he was accused of having connived at De Lorenzo's conspiracy.[1]

The Chamber was dissolved in February 1963 and the general election fixed for the end of April. In March Pope John received in private audience Khrushchev's daughter and journalist son-in-law, something that would a very short time before have seemed unthinkable. It is said that the peasant Pope asked his Russian peasant visitor – the wife, Rada – to tell him the names of her children because, as he said, a woman's eyes light up when she speaks of them. On 10 April the Pope published a major encyclical beginning with the words *Pacem in terris*. It was addressed to 'all men of goodwill' and it emphasised the conception of freedom of conscience; finally it supported the idea of collaboration, in the cause of peace and social justice, between men of differing persuasions. Thus the encyclical displayed a threefold revolution in the attitude of the Vatican. The Pope

[1] See below, p. 59.

meant to support the opening to the Left but what he probably
brought about was that many working women, the wives of the
Red Belt Communists for instance, now felt free to vote Communist too. There had been unprecedented prosperity but now
this was cracking up: in any case it has been seen that the effects
were not such as to diminish envy or other causes of friction.
Television had arrived in Italy in 1954. By 1960, when the
Tribuna politica or political debate was introduced, the Communists were represented on it. By this time television was ceasing
to be a luxury and the Communist speakers succeeded in being
particularly effective. In August 1962 in his influential weekly
review *Rinascità* Togliatti had written an article of great moderation which seemed to accept the *Centro-Sinistra* experiment as
almost commendable. But in the spring of 1963 the Communists
hurled the old abuse, angrier now, against Nenni for betraying
the working class, seceding to the imperialists and so on.

Thus it was no surprise when the Communists gained over
a million votes in the general election of April 1963. It was more
remarkable that the Social Democrats gained noticeably, and
even Nenni's party gained three seats in a slightly larger chamber
(630 instead of 596 seats). Interestingly enough, the Liberal Party
gained nearly as much as the Communists did: its parliamentary
representation more than doubled. No doubt the Liberals won
a little from the Monarchists who all but disappeared. But they
must have gained on a large scale from the Christian Democrat
Party which lost nearly three-quarters of a million votes. Nevertheless in the new Chamber, although there would be 166 Communists representing just over a quarter of the electorate, there
would still be 260 Christian Democrat deputies. The idea of the
opening to the Left was by no means defeated. It had been conceived as a way to disarm Communism. If it had signally failed
to do this, how was it to proceed? Whispers of bringing the Communists in from outside were already to be heard, though not so
distinctly as they would be five years later.

6 Communists and Christian Democrats after the General Election of April 1963

In many ways the Communist Party of Italy seemed the strongest force in the country after the election of 1963, the most united, the best organised, seemingly invincible. It was the usurping inheritor of the Italian Socialist tradition. The Socialist Party of Italy was founded in 1892 before any large-scale industrialisation had affected the country. When the motor-car and the Breda steel and Pirelli's rubber industries sprang into vigour early in the twentieth century a party of Socialist intellectuals was ready to provide the new industrial workers with an ideology. The most remarkable thing, however, about the Italian Socialist Party was that it rapidly acquired an important agrarian following, not so much among the poor and landless peasants, who were often illiterate and priest-ridden, but among the better-off, sometimes indeed fairly prosperous, *mezzadri* of central Italy. To this day the less well-off peasants of the Veneto or in the south vote Catholic, that is Christian Democrat, but the richer peasants of the former Papal States and Tuscany often voted Socialist as soon as they could vote at all, that is from 1913 onwards – the only obvious reason for this was their inherited desire to vote as emphatically as possible against the Church. The rise of the Communist Party threatened to split the Red vote of central Italy. After the end of Fascism, when women were enfranchised, families were often divided because many wives of Socialist or Communist husbands voted Christian Democrat as their priests made clear to them was their duty: otherwise their children would burn in hell for many generations, they were told. After the Socialist split in 1947 and the elections of 1948 the Communist Party steadily consolidated its hold on central Italy. Since the local elections of 1946 many communes there had been ruled by a Socialist–Communist coalition. As the years passed, local government in the Red Belt was more and more controlled by the Communists: to a great extent this became a self-generating phenomenon as the Red mayors and communal councils extended their influence and in consequence their patronage. The post-war Communist Mayor of Bologna, Dozza, really a moderate Socialist, remained in office for many years and was indeed a local baron and a worthy one. Although in other areas of Italy the peasants more usually voted Christian Democrat, in critical periods the Communists picked up votes in the south. Obviously the industrial workers provided the backbone of the Communist Party throughout the period. The Communist Party in fact exerted great influence and power through the trade unions. In Bari in 1944 the CGIL had been founded to include all types of trade union, and in June 1947 it held its first congress, five months after Saragat's break with Nenni. In 1948 non-Communist trade union members began to break out of the CGIL to form their own Christian Democrat or Social Democrat unions – there was even a small neo-Fascist one quite soon. Nenni's followers, however, remained in the CGIL, becoming increasingly dependent on Communist leadership and often, like Lama, indistinguishable from the Communists. Thus the average member of a trade union came automatically to adopt a Communist attitude, to accept Communist slogans. There was a network of co-operatives controlled by the CGIL too. Both unions and co-operatives were increasingly vehicles of Communist propaganda; they achieved much less in terms of labour rights and co-operative business. To this day a striker in Italy gets no regular strike pay and no National Assistance; hence the brevity of Italian strikes.

In the middle fifties before the effects of the boom were felt, better-off workers like those at Fiat sometimes dropped their Communist allegiance. In April 1955 only 59 members of the CGIL were elected as shop stewards there as against 133 members of the Christian Democrat, Social Democrat and small Fascist unions. A few months later, however, the CGIL did very well in Pirelli's rubber factories and those of Alfa-Romeo which was controlled by IRI. Once the boom was effective it seems clear that it tended to help the Communist Party. The hope that less unemployment and better earnings would undermine

Communism in Italy proved deceptive. The increase of prosperity brought above all an increase in the number of television sets, and the Communists succeeded in presenting better 'images' and simpler, more convincing arguments – or certainly slogans – than anyone else. This was due to their astonishingly good organisation of propaganda. The Communist Party line seemed clear and consistent, however unsound and insincere, and was presented in every imaginable form. The fact that the boom worked erratically, convulsively and explosively – it has been seen that it was not true to claim it as totally unplanned – was easy to exploit in terms of discontent. There was always a majority of people which could be made to feel that others had profited more from economic expansion than they had. The wishes of the great block of Communist voters in central Italy were often at most what had once been called Socialist reformist, i.e. moderate labour in English terms. But they were under-represented in the party machine because a moderate attitude worked less well as propaganda. The Communist leaders stimulated discontent among their members in central Italy by decrying the *mezzadria* system of sharing crops; already at this stage a law had increased the peasants' share of the profits from 50 to 58 per cent in central Italy – in the south it had often been much less than 50 per cent.

An important part of Communist power in Italy lay in the Communist press as well as in the organisation of what was shown on television. The Communist daily, *Unità*, was read like a bible by simple, even not so simple, people. The party's weekly review, *Rinascità*, was aimed at the educated classes. In addition the Communist Party had links with the most outstanding intellectuals. Apart from those, like the painter Guttoso, who belonged to the party, most of the influential writers, Moravia for one, although not orthodox members, favoured the Communist approach. In particular some of the leading publishing houses were on the side of Communism if it came to it; there was Einaudi in Turin owned by the President's son, Giulio, who had become Communist in Fascist days; there was Feltrinelli in Milan who much more recently espoused the anarchist cause. This meant that books and reviews inclined in the Communist direction as expressed by the writers, towards an ill-defined state-socialism and for many years towards sympathy with Moscow against Washington. Other pockets of influence helped the Communists who, unlike the

Socialists, always succeeded in profiting from leftist sympathies. At one time a group of intellectuals in the *Banca Commerciale* in Rome and Milan (owned by IRI) played a part in the intellectual world because they were intelligent and imaginative and also because one of them was married to Croce's eldest daughter whose salon attracted gifted young historians. Here a review was published called *Lo Spettatore Italiano* which insisted that 'our' Communists were 'different' and could not be kept at the door since they were an integral part of Italian society. Elena Croce's husband at that time was Raimondo Craveri and the director of the *Banca Commerciale* was Rafaello Mattioli; a junior colleague called Franco Rodano was a Catholic Communist excommunicated by Pope Pius XII.

Of course Khrushchev's exposure of Stalin and the anti-Russian rising in Hungary caused a crisis in 1956. But it has been seen that the Italian Communist leaders had always let it be whispered, especially into the ears of intellectuals, that they were 'different', and Togliatti succeeded in weathering the storm with only a few defections. For the rest of his life he was able to persuade the Russians that he was loyal to them while at the same time expounding theories of 'polycentrism', i.e. that the Italian Communist Party would decide for itself what Communism was to mean in Italy as any other national party might.

The revelations of 1956 at last shook Nenni profoundly and made his co-operation with some kind of leftist Christian Democrats, already mooted, into a remote possibility. But Nenni had shown himself no organiser, and the Italian Socialist Party had neglected modern techniques of power. The Party organ, *Avanti!*, named long ago after its German model, had retained a surprising degree of inherited prestige, but the PSI had nothing else. Its obvious weakness seemed to nourish the factiousness of its leaders, and this in its turn worked to Communist advantage. The elections of 1958, in spite of Hungary, did not show an important weakening of the PCI, and in 1963 it has been seen that it gained decisively. All the time the PCI had helped to block greatly needed legislation by opposing and condemning whatever the government of the day proposed. Then, because legislation was blocked by Communist opposition – the Party was enviably free of all governmental responsibility in Rome – the Communists abused the Government for its paralysis. All politicians other than themselves were corrupt, they implied, and people believed them.

It should be added, for it is not irrelevant, that where the Communists were responsible for local government in central Italy they were not quite as efficient as the legends they circulated about themselves. In Bologna, and in other cities they controlled, they too were bureaucratic and slow,[1] although one would have thought they would have been concerned to display their competence and that they would have been on their mettle in the face of the prefect who represented the central government in each provincial capital.

It has been claimed, probably with justice, that the Communist Party, although increasingly middle class and often a family affair, maintained its hold largely because it provided simple working men with a status they could not otherwise acquire. In other parties, it was more difficult for a worker to become a party official. As for the Socialist Party, its party machine merely creaked; it failed to offer the myriad 'little' positions in a big bureaucracy.

Indeed, as Giorgio Galli of the *Mulino* has emphasised, post-1945 Italy increasingly developed a bad two-party system in which, however, one party chose to be in apparently permanent opposition. At the same time the Communists condoned much legislation by commission which their veto could have prevented or at any rate delayed. On the occasions when they might have seized power, the Communists made it obvious that they had no real wish to do so. As for suggestions that they might enter the Government coalition, although they complained of being excluded from it is difficult to believe that they were prepared to share governmental responsibility before 1970 at the earliest.

The other party in this two-party system, based on democratic rules in which neither party altogether believed, helped the Communists by its divisions. Indeed the Christian Democrats, after De Gasperi's day, became as factious as the Socialists. There was not only the division between anticlerical Christian Democrats like De Gasperi himself and the Catholic hierarchy – it has been pointed out accurately that a leader like De Gasperi depended upon the Church for much of his voting support but was then prevented from legislating as he thought right by the innate con-

[1] See Giorgio Galli, *Il Bipartitismo Imperfetto* (1966).

servatism of the Church. This was true of nearly all his successors except for the short period from the emergence of the political influence of John XXIII until the eclipse of that influence within the Church after the second Vatican Council dispersed.

What other reasons explain the friction between the Christian Democrat *correnti*? Although the Church laid down certain dogmas it did not completely dictate a man's political opinions to him as the Communist Party did. Economic change loosened up the Catholic view at much the same time as John XXIII and the second Vatican Council from 1962 to 1965 did so. Although Paul VI then tried to retreat, many Christian Democrats would never again support this attempt. Economic change was a little easier for the Communists to absorb and interpret as it suited them, without more than a few pro-Chinese, usually very young people, breaking away. The Christian Democrats had a larger proportion of graduate members, more arts graduates in particular and fewer technicians than the Communists: unlike France, Italy's schoolteachers were mostly keen Catholics. All these people tended to be talkative and disputatious, nursing the Italian passion for playing with political ideas.

In so spacious a party the natural dissension between the Right, in the sense of the champions of property and the *status quo*, and the Left in the sense of the champions of the redistribution of wealth in the interests of the poor, was deep. The rightist groupings did not produce so much faction as the others. Among Scelba, Pella and their friends – Pella in particular was closely associated with some of the big industrialists – there was strong distaste for social reform. They avoided confrontations with Catholic Action of which, however, their colleague, Gonella, had earlier been a spokesman in the organ of the Vatican, the *Osservatore Romano*.

Catholic Action, for many years dominated by Luigi Gedda, was a rigidly Catholic organisation of laymen, which had defended the Church against the assaults of Fascism. Among its many organisations the *Associazione Cristiana di Lavoratori Italiani* or ACLI united the more devout Catholic working men. When, however, the more politically-minded Catholic trade unionists broke away from the CGIL they founded their own *Confederazione Italiana di Sindacati Liberi* or CISL which paid no homage to Gedda's authority and was on bad terms with ACLI. Prominent in the CISL world was Giulio Pastore who gathered around him leftist Catholics in a group called the *Sinistra di Base*.

Two other men originally took the initiative on the left wing of the Christian Democrat Party, Dossetti and Fanfani. It has been seen that these two had been fellow lodgers at the *Comunità del porcellino* at the end of the war.[1] For different reasons they wished to act independently of industry and economic power in general, but also to some extent independently of their own Church. While Pius XII was Pope he supported Catholic Action and Gedda, its chief, in resisting the plans of these leftists. Dossetti surrendered and became a priest in time to help prepare, in conjunction with Cardinal Lercaro of Bologna, John XXIII's ecumenical council. Fanfani, however, fought on fitfully until the spring of 1963 and of course beyond that: he kept their joint *Iniziativa Democratica* going until his temporary collapse in 1959.

The stormy character of this strange man, his alarming background and intelligence, made him into what seemed like the evil genius of Christian Democracy and the generator of its factions. It has been said that his aims were never clear, 'his means seemed too harsh and Machiavellian, the man himself too thirsty for power. While flattering everyone, he caused everyone concern; the hierarchy understood that he intended to remove himself from its control and refused to be bought off with clerical rhetoric from him. Such rhetoric, on the other hand, discouraged the most open minds both in the Party and in the Catholic world.'[2] He promoted 'sharp' opportunists rather than idealists in his many appointments. Fanfani claimed to be the pivot of the party, heir, according to an alleged 'last message', to De Gasperi. In *Iniziativa Democratica* he had had close to him Rumor, Colombo and Moro, all involved with him at an early point in or under the Ministry of Agriculture. Curiously, they, who later formed the moderate group of the *Dorotei*, were for a time more closely linked with the Church than with the Christian Democrat Party. Where would the exit be from this labyrinth? To be a Communist was child's play by comparison with being a Christian Democrat.

Whether genuine idealism was woven into Fanfani's character

---

[1] Carlo Falconi, *Pope John and his Council* (1964), p. 78, and see above p. 22.

[2] Galli, p. 213.

with the rest or not, his friends believed it to be so. The strangest of them all believed it most profoundly. This was Giorgio La Pira, the Sicilian who became Florentine and in 1961 Mayor of Florence. La Pira's Catholic devotion was medievally mystical; he behaved like a new Messiah who could make all men love one another and he believed in himself as such. But he did not, he could not, heal the divisions in the Christian Democrat Party; he only added another one.

It would be ultra-Italian to leave Christian Democracy in this condition of disintegration. The trio Colombo, Moro, Rumor had positive attributes, attributes which could lead to positive action and co-operation with other parties. If the Communists still refused to accept responsibility, these men did not. Moro was a tortuous conciliator, Colombo before everything an able economist. Mariano Rumor was originally thought by his friends to be the most clear-cut personality of the three. He was a devout Catholic from the Catholic Veneto, from Vicenza to be precise. He had been a schoolmaster and he had worked closely with ACLI in the Veneto. He had impressed De Gasperi as a young man and had worked under Fanfani at the Ministry of Agriculture when land reform was launched. He was upright, industrious and a convinced reformer in terms of Catholicism, of a Catholic Christian's duty. On one occasion Malagodi attacked Rumor along traditional lines. Planning, he said, was a very old story. As for the regions, of which Rumor was a champion precisely as a necessary part of planning, they were part of the Catholic vendetta against the lay state united by the *Risorgimento* against the power of the Church. Rumor replied sharply that the Liberal state had handed down too much disorder to the republic; the regions were not aimed against Italy's unity, but comprised a 'more rational and modern way of expressing it'.[1] For a time Rumor seemed to be Christian Democracy's clearest spokesman and perhaps its most faithful servant. Early in 1964 he succeeded Moro as Secretary-General of the Christian Democrat Party and remained in this position for another four years.

For men like Moro, Colombo and Rumor the meetings of the twenty-first General or Second Vatican Council convened by Pope John XXIII were of fundamental importance. The Pope began to define the work of the Council already in November

[1] See G. Ghirotti, *Mariano Rumor* (1970).

1960, and in 1961 he named Cardinal Bea to be head of the Secretariat for Christian Union. It was evident that John XXIII was appealing to 'all men of goodwill', as he said in the encyclical, *Pacem in Terris*, to make Christian unity something true: in a way he was appealing from the Curia and the Italian episcopate to the outside world. At last the first session of the Council opened on 11 October 1962 towards the end of Fanfani's administration which nationalised electricity. Mattei's death took place towards the end of the same month; the Council's first session lasted until 8 December. On 20 November a majority of the assembled fathers rejected a draft scheme put before them on the sources of revelation, and it became clear that many of the non-Italian bishops wished for a more modern Catholic Church. This was an important factor, strengthening the pressure towards the opening to the Left in Italy. It has been seen that in the spring of 1963 John XXIII made gestures in the direction of Moscow from which the Italian Communists profited in the general election. But then in June 1963 John XXIII died and the second session of the Vatican Council, from 29 September to 4 December 1963, was held under his successor, Cardinal Montini of Milan, who became Paul VI. The Fathers of the Church expressed their approval of more frequent use of the vernacular, greater lay participation and the development of the ecumenical idea. Paul VI, apart from the foreign travel he undertook, was a more conventional Pope, modern only to the extent that Catholic opinion might now demand.

There can be little doubt that Moro and Rumor continued to feel the influence of John XXIII although they knew that his attitude towards Communism had been dangerously naïve. Paul VI was not without influence over the leaders of the Christian Democrat Party. However, the third and fourth sessions of the Vatican Council, in the autumn of 1964 and 1965 respectively, brought nothing of general importance.

It has been seen that Togliatti was intensely aware of the possibility of using Catholic feeling for Communist purposes, and there is no doubt that John XXIII and the first session of the Vatican Council had impressed him and confirmed his hopes. Already on 20 March 1963, three weeks before the encyclical *Pacem in Terris*, Togliatti had made a speech at Bergamo in which he suggested discussions, even understandings with Catholics who could not be indifferent, he said, to the international dangers and economic problems of the world. Immediately after Togliatti's death in August 1964 his party announced that it would be the duty of his political heirs to develop his assumption that a religious conscience could contribute to the social aims of Communism, since religion need not necessarily be allied with the 'exploiters' and 'monopolists'. Thus a 'dialogue' with the Catholics must be pursued. At the end of 1964 a Catholic named Mario Gozzini published a book called *Il Dialogo alla prova* with contributions from five Communists and five Catholics. The best known among these Communists was a deputy called Lucio Lombardo Radice, who insisted that although Catholicism had a 'reactionary' past there was now no reason why it should not become revolutionary; above all it should banish its traditional fear of Communism. The well-known Jesuit, Giuseppe De Rosa, on the staff of the Catholic journal *La Civiltà Cattolica*, commented upon Gozzini's publication that it had mainly shown that the Communists were trying to use the whole 'dialogue' as propaganda for their own political purpose and there was little reason to believe that they had really been converted to a belief in any kind of liberty.[1] Articles showing how much Catholics had in common with Communists nevertheless continued to be published in *Unità*, and especially in *Rinascità*, during 1965.

[1] See G. De Rosa, *Cattolici e Comunisti oggi in Italia* (1966).

7 Italy under Moro

When Fanfani resigned after the election of April 1963 Segni certainly did not wish to reappoint him as Prime Minister. The Christian Democrat Secretary-General, Aldo Moro, a member of the *corrente* of the *Dorotei*, then cautiously leftist, seemed the man to choose as Fanfani's successor. Moro came to terms, as they both thought, with Nenni. But on the night of San Gregorio on 14 June 1963 Riccardo Lombardi staged one of his *coups* from within the PSI. He was said to be in league with Fanfani – what a couple they were – in upbraiding Nenni for having agreed to accept too little on planning, the regions and possible co-operation with the Communists in central Italy where the regions would be Red.

This spelt deadlock until the PSI congress due to meet in October 1963. Until then Segni appointed a caretaker *monocolore* or one-party Government headed by Giovanni Leone, President of the Chamber. Perhaps the most interesting thing about this Government was that Rumor became its Minister of the Interior. He immediately took measures against the members of the Mafia in Sicily, refusing all attempts to deflect him. Having visited Sicily in person he went north to deal with terrorism in the South Tirol.

At last in December 1963 Aldo Moro succeeded in launching a *Centro-Sinistra* Government which actually contained members of the PSI. Indeed Nenni himself became Deputy Prime Minister, and this time the Minister of the Budget and of Planning was Antonio Giolitti (grandson of the famous Prime Minister[1]), who had been a Communist until 1957. This new coalition expressed a revolutionary change, for Nenni had insisted upon the acceptance of planning as a principle, as a 'myth' in the non-English sense of inspiration. His surrender was rather on foreign policy since he had now accepted Italy's obligations under NATO provided they

[1] The years 1900–15 are usually referred to as the age of Giolitti in Italy.

were interpreted as strictly defensive and limited to Europe. This acceptance, which spelt a serious breach with the Communists, caused a new Socialist split. Lombardi, protesting as ever, remained in Nenni's party; indeed he was promoted to become editor of *Avanti!* for a short time. However, a group of twenty-seven deputies led by Vecchietti and Lelio Basso broke away and founded their own 'Italian Socialist Party of Proletarian Unity' (PSIUP) which has ever since taken up a position rather to the left of the Communists.

In January 1964 the new Government announced a Five-Year Plan to which the objections of the right-wing Christian Democrats were hard to conceal. The plan involved a lengthy and ambitious statement made more complicated by the fact that the Italian Government was in fact bound to try to plan for at least two types of society: developed and backward. The critical issue of how binding the Plan was to be was evaded with the kind of skill for which Moro was to become proverbial. By dint of avoiding precise definition he succeeded in remaining Prime Minister for nearly five years with a slight variety of colleagues. Although little tangible was achieved in this period, a very slight leftward tendency was generally maintained which was to the country's advantage. But real planning could not hope to succeed without better-educated civil servants, and, in spite of much talk and some effort, education in Italy improved very little: a school set up at Caserta for future civil servants was not very efficient. Any slight beneficial reform was always overtaken by the increase in population, and particularly by southern emigration to Rome and the north. Moro did in fact resign on 26 June 1964 over a matter of subsidising private, i.e. Catholic, schools, but was back in power within a month. That July was stormy, with Pacciardi inveighing against the tyranny of the parties at a meeting in Bari and a certain General De Lorenzo playing with the idea of a *coup d'état*.

The year 1964 was marked politically by two major casualties. On 7 August Segni was struck down by cerebral thrombosis, not mortal but incapacitating. In his place late in the year Giuseppe Saragat was elected to be President. The intrigues surrounding his election were intricate and embarrassing. But the result was a useful one. Saragat seemed to have the stature and the experience, and he was a warm advocate of the opening to the Left.

Shortly after Segni's stroke, on 23 August Togliatti, now aged seventy-one, died on a visit to Russia. His successor as Secretary-General to the Communist Party was Luigi Longo, who had been a successful partisan leader twenty years earlier, but who, if humanly pleasanter, was intellectually nowhere approaching Togliatti. Truer heir to the latter was Pietro Ingrao, who became leader of the PCI in the Chamber. A new period opened in the leadership of the Italian Communist Party whose divisions now became more apparent. Local elections in the autumn of 1964 slightly increased support for the party but the spirit of *contestazione* or challenge began to emerge: the challengers to Communist and Catholic authority tended to converge. Pope Paul VI was gradually discovered for what he was, no heir to John XXIII but a vacillating, even backward-looking Pope, who impelled eager young Catholic reformers further into their *Dissenso*, their protest against the cautious reassertion of Catholic conservatism.

When Saragat was promoted to the Quirinal from the Farnesina, that lovely palace to which the Italian Foreign Office had been moved from the Palazzo Chigi, Moro took the momentous step of bringing back Fanfani to be Foreign Minister. What did he inherit as Italian foreign policy and what would he bequeath to Italy? Since Sforza there had not been a clearly formulated policy. Italy continued to be strongly European, active at Strasbourg as well as at Brussels and in 1958 host country for the EEC treaty signed at Rome. Until the *Apertura a Sinistra* she was a keen supporter of the Atlantic policy (though one wonders about Fanfani's earlier period as Foreign Minister); she had received much economic help from the United States and there was a large Italian-speaking population there of which the Italians in Europe were intensely aware. The activities of Clare Luce, however, who was American ambassador in Rome for several years, did not contribute to Italian–American friendship; about Mrs Luce's interference, Communists, Socialists and supporters of Mattei were bitter. But the Russians did little to satisfy the Italians except at last in 1955 to drop their veto to Italy's entry into the United Nations in a bargain which also brought in several Russian satellites.

In 1955 also, the Austrian State Treaty was signed and the occupying powers withdrew from an Austria which was declared permanently neutral. This re-created serious trouble in the South Tirol or Alto Adige, which, in conjunction with the Trentino, had been made into an autonomous region years earlier. South Tirol itself, where the towns of Bolzano and Merano were predominantly Italian owing to inter-war immigration, had a two-thirds German-speaking majority. The tradition among these people, many of whom had opted for Hitler's Germany between 1940 and 1943, was a near-Nazi one; the vile way in which they had been treated by the Fascists had encouraged their racial fanaticism. Tirol with Styria and Carinthia had long nursed feelings of this kind against their Italian or Slovene co-citizens, and the South Tirolese were encouraged from Innsbruck. They protested indignantly at their incorporation in Trentino–Alto Adige which as a region had an Italian majority to whose decisions the South Tirolese were thus subjected. The Italians felt that they had been very generous to have allowed the pro-German South Tirolese optants to return, and they felt they had more pressing tasks than to teach the Italian railwaymen in the Alto Adige to use the German place-names there. Certainly the regional officials were nearly all Italian and in the courts it was risky to put one's case in German because Italian judges might not understand. On the other hand it really was true that German-speaking people seldom applied for regional posts because they thought they would be isolated. In the schools the German-speaking children were taught in German but the courses were too greatly determined by the Ministry of Instruction in Rome. In the later fifties and early sixties many South Tirolese Germans were obviously in touch, not only with Innsbruck, but also with Nazi-flavoured centres run by East German refugees in Munich. In the sixties some of the South Tirolese took to terrorism. It was a very disagreeable situation. According to an agreement made by De Gasperi in September 1946 with the Austrian Foreign Minister Karl Gruber and incorporated in the Italian peace treaty, the Italians had recognised Austria's interest in the matter, i.e. that the South Tirol was not a purely internal Italian question; this had been a strange indiscretion on the part of De Gasperi, the former Austrian. So Austria appealed to the United Nations in 1959 and 1960. The Italians were able to nullify this action because they could point out privately to their European partners that to encourage the South Tirolese would encourage the worst kind of intransigence in Germany. In 1964 Italy began direct negotiations with Austria about the South Tirol. An Italian Commission of Nineteen chaired by the Socialist, Paolo Rossi, had worked since 1961 on the possibilities of a compromise which, after seemingly unending delays often caused by a change of government in Rome, arrived at a result which was accepted in principle by the South Tirolese in 1969. The acceptance was partly due to a tardy recognition of the damage done to tourism by South Tirolese acts of terrorism. Finally, on 5 December 1969, the Italian Parliament approved an Italo-Austrian agreement accepted by the Austrian National Council ten days later. According to this the legislative power of the province of Bolzano (or Bozen) within the region of Trentino–Alto Adige was extended. It was stated that this was an internal Italian question, but also that any future differences of opinion about the enactment of the De Gasperi–Gruber agreement should be submitted to the International Court of Justice at The Hague.

The American administration, when John Kennedy was President, smiled on the Centre-Left experiment: with Saragat at the Foreign Office all should have been well. Fanfani, however, was another matter, particularly when it was rankling with him that he had not been elected President of Italy – there would certainly be seven years to wait. It was an extraordinary thing for this man to find himself again responsible for Italian foreign policy. He was opposed by definition to the attitude and policy of the United States. He would have liked to be a leftist de Gaulle. In any case Italy tended to oppose France in the Common Market. For Italy was bound to wish for British membership in order to reduce the preponderance of France and the German Federal Republic, particularly since those two powers had emphasised the Paris–Bonn axis.

In September 1965 Fanfani was elected President of the General Assembly of the United Nations for its twentieth session, 1965–6. What can have been the thoughts of this formerly racialist, half-Tuscan, half-Calabrian, as he presided over this by now predominantly Afro-Asian society? Did his thoughts go back at all to the racialist views he had held in his youth in Milan? Since the time when he himself had started government by the *Centro-Sinistra* the relations between the Farnesina and the State Department had obviously become more difficult. The United

States had established missile-bases in Italy in accordance with NATO undertakings: these were dismantled in 1963 and their place taken by submarines carrying Polaris missiles not based on Italy. Nenni in 1963 jibbed at the whole idea of a multilateral nuclear force but by 1965 this plan was being dropped, though Vietnam had become a major issue. Typically, Fanfani, when President of the U.N. Assembly, chose to transmit a message from La Pira to the American Government about some Vietnamese with whom La Pira had recently spoken: La Pira had done this soon after he had ceased to be Mayor of Florence with a *giunta* or town council that had been partly Communist. The Socialists with Nenni had been brought into Moro's Cabinet on the understanding that they did not govern locally with Communists; at the time those who dissented had joined the Socialist Party of Proletarian Unity. La Pira, however, was a law unto himself – in Florence he was revered as a saint or deplored as a madman. It was from him that the Italian Foreign Minister chose to deliver a message to the State Department. It was to have been a private message but of course it did not remain so: indeed La Pira himself published an interview about it. There was nothing for Fanfani to do but to resign and this helped of course to bring down in January 1966 Moro's most recent and precarious government formed in March 1965. Yet in his subsequent administration Moro reappointed Fanfani who remained at the Farnesina as long as Moro was Prime Minister, that is until 1968: this hardly contributed to a coherent foreign policy, a stable government or a united Christian Democrat Party.

The years 1965 and 1966 saw an endless discussion of planning, of reforms, which were bound to be expensive, versus the need for investment. In 1965 legislation was introduced gradually to destroy the system of the *mezzadria* which was felt by the Left to be degrading for the peasants who shared crops. Its effects do not appear to have prevented any peasant from leaving the land. 1966 was a year of disaster for Italy, the collapse of a number of houses in Agrigento in Sicily in August,[1] followed by the appalling floods in Tuscany and the Veneto in November. Italy is subject to natural disasters of this kind, and Florence had been periodically flooded. Horace Walpole writing to Richard West from Florence in November 1740 said 'Yesterday, with violent rains,

[1] See below, p. 108.

there came flouncing down from the mountains such a flood that it floated the whole city. The jewellers on the Old Bridge removed their commodities, and in two hours after the bridge was cracked', and he expected terrible news from Pisa. In 1966 the modern use of oil had greatly enhanced the catastrophe.

The whole conception of planning seemed mortally wounded by the floods, and Moro called upon the Italians to prepare for austerity in order that the Government should spend what it must upon relief and restoration. La Malfa expressed his disappointment that Moro did not impose a tax on wealth, steeply graded, as a gesture in favour of the principles of social justice. One measure taken by Moro in order to have more money for relief and rehabilitation was that the high contribution – 83 per cent of the whole cost – paid by all employers towards social insurance, which had been reduced by the state taking over more of the expense in 1964, was now reimposed. In fact less money than had been foreseen was officially invested in the next Five-Year Plan just accepted. The extraordinary thing was that the Italian economy was not seriously injured by the floods: the upward trend, which had returned in 1964, continued with only a minor disturbance until 1969.

With the death of Togliatti, the fall of Khrushchev and with the Centre-Left Government, including Nenni as Deputy Premier, in its beginnings, the Italian Communists had a new feeling of isolation. Longo, Togliatti's successor as Secretary-General, was a lightweight, already too old, and was soon weakened by a stroke. Within the party something of a feud had arisen between Giorgio Amendola and Pietro Ingrao and their respective supporters. Steeped in party jargon though he might be, Amendola, like the traditional Socialist he at bottom was, thought in terms of a broad Communist–Socialist alliance on the Left. He represented the Red Belt Communists who hated the breach with Nenni. Ingrao was a younger man who was able to pose with some justification as Togliatti's heir. He favoured an approach to the leftist Catholics, as it were cutting out the Socialists. The Red Belt Communists, incidentally, were close to Dossetti and on good terms with their 'progressive' cardinal at Bologna, Lercao. After the end of 1964 Ingrao was obliged to recant by stages, until his surrender at the Eleventh Congress of the Communist Party in January 1966 was made complete. It is not quite clear why Lombardo Radice was allowed on the same occasion to express his enthusiasm for diversity of opinion.

At this Congress Longo read out the usual mammoth speech in which, after referring to Togliatti's appeals to the Catholic world to co-operate in the fight against the American imperialists, he made the following statement:

We consider that religious peace could be safeguarded by, above all, a solid contribution to the development of socialist society which should favour the loyal and fruitful participation of all believers in the construction of a society freed from exploitation. Clearly we advocate a state which is a completely lay one in practice. As we are opposed to a confessional state, so we are against the atheism of the state. Thus we are opposed to the state granting any privilege whatever to any ideology or philosophy or religious faith or cultural or artistic trend, at the expense of others.

This was as far as the official Communist Party of Italy would go for the time being, and it was quite a long way. From the Christian Democrat side Rumor, for instance, when winding up at the party assembly at Sorrento on 3 November 1965, had said that, where the Church could parley with heretics as individuals, the Christian Democrat Party must say no to the suggestions of a dialogue with the Communists. The idea of the Communist–Catholic *Dialogo* was, however, popular with the young, although people like De Rosa greatly feared the exploitation of their enthusiasm by the Communists, and Rumor said to Ingrao in the Chamber: 'Surely we have been carrying on this debate for the last twenty years.'

At Sorrento in November 1965 Rumor made it clear that while bewaring of the Communists, he hoped for the reunification of the Socialist Party in the near future. In spite of many difficulties it had been possible to work with Nenni as Vice-Premier in Moro's Cabinet. Clearly Nenni was not trying to swallow up the Christian Democrats in a Socialist revolution. Indeed, he had become so unrevolutionary that his Christian Democrat colleagues felt that he would be more useful if he were backed by a bigger Socialist Party which might draw away some of those who had been voting Communist. In fact a Socialist congress opened in Rome on 10 November 1965 to discuss reunification between

Nenni's PSI and the Social Democratic Party at this time led by
Tanassi. Its senior chief, Saragat, had now been President of the
Republic for nearly a year. He did all he could towards reunification and was already considered to be interfering more than his
constitutional position warranted; in his own view he remained
sufficiently in the background.

The Socialist Congress at Rome in November 1965 was unsatisfactory. The Social Democrats pressed for reunification and
Nenni was warmly applauded when he insisted that the Socialist
Party must be genuinely independent of the Communists. It was
not quite easy to forget how much his attitude had changed, and
there were unkind whispers suggesting that he wanted to be able
to claim more spoils of office, more jobs in the *sotto-governo* or
substratum of officials attached to any administration. The other
speakers for Nenni's PSI, however, were unhelpful. Nenni's
successor as Secretary-General of the Party, De Martino, insisted
that the Socialists must still be free to join with the Communists
in local government as they did in the Red Belt, and, rather
provocatively, he claimed that Communists should be eligible for
the European Parliament. The former Communist, Antonio
Giolitti, expressed his impatience with the working of the *Centro-
Sinistra* which he thought not sufficiently 'incisive' or 'advanced',
and Riccardo Lombardi was as unblushingly disruptive as ever.
So reunification was postponed again.

In June 1966 local elections provided a new impetus. The
Socialists lost but the Social Democrats, such as they were, had
a big success; the Christian Democrats gained a little while the
Communists remained steady. Further steps to reunify the
Socialists were not, however, taken until the autumn. Nenni made
speeches positively attacking the Communists who, he said, had
condemned themselves to *immobilismo*, the paralysis of which the
Centro-Sinistra was increasingly accused. The Communist Party,
he asserted, had done nothing to justify its claim to champion
political liberty which he now said was no less important than the
socialisation of the means of production. The thirty-seventh
congress of the PSI opened on 27 October 1966. On 29 October
Nenni said that the independence and initiative of the Socialist
Party must be restored, and De Martino, implying the same thing,
said that the PSI must again take up the position of being the
third big party; it was in fact De Martino whose attitude had

changed. Even Giolitti seemed prepared, pleading that 'neo-capitalistic' well-being should be based not only on equality and fraternity but also on liberty. At last with manifold applause on 30 October the reunion of the Socialists with the Social Democrats was declared after a divorce which had lasted very nearly twenty years. Adjacent offices were installed in the Corso in Rome and each side provided an editor for *Avanti!* In fact the reunion was quite unconvincing except to Nenni – only he in the PSI had really changed. The floods in Tuscany and Venetia towards the end of November obliterated all other problems until the end of the year.

8 *Dialogo* and *Dissenso*

The atmosphere in Italy in 1967 was uncertain and uneasy. The Six Days War between Israel and the Arabs was something close to Italy, not only in geographical terms. The policy of Mattei had involved Italy with the Arabs, and the reactions of the Farnesina did not contrast very sharply with the pro-Arab cries of the Communists. At the Christian Democrat Congress in October, however, the Secretary-General, Rumor, insisted that Italy's allegiance to the Atlantic Pact must not waver; at the same time he made intelligent reference to the ferment among the young.

Terrorists, probably based on Austria and connected with Norbert Burger's near-Nazi group, caused the death of four Italian soldiers at Forcella in the Alto Adige early in June; Burger's trial at Linz ended in his acquittal. At the end of September two Italian policemen were killed at Trent by a bomb they were trying to remove. It is not certain who had placed this bomb, for Italian anarchists and 'pro-Chinese' or 'Maoisti' were already contemplating violent action too. The quarrel between Italy and Austria, it has been seen, was about to be discarded as out of date – Forcella was almost a postscript.

In April 1967 General De Lorenzo had been dismissed from his position in the Army as Chief of Staff: he had been accused in the *Espresso* by a journalist called Scalfari of using the Sifar or Military Intelligence to plan a *coup d'état* against Moro's Government in the summer of 1964, the year when Segni had collapsed as President: it was even said that Segni had been implicated. In December 1967 a case was brought against De Lorenzo, who was, however, exonerated in the following July. Scalfari had by now been condemned to seventeen months in prison for attacking De Lorenzo, so, with a general election pending in May 1968, he was adopted as a candidate of the PSI in Milan to save him, should he be elected, from serving his sentence. In fact De Lorenzo was elected a Monarchist Deputy.

The De Lorenzo affair embittered Italian politics for years; it was not until November 1970 that a court in Rome pronounced him to have been guilty of illegal aims.

In the middle of February 1968 the regions, which everyone at all on the Left, people like Rumor included, felt could contribute so much to the better organisation of Italian society, were voted into geographical existence by the Senate: this had already been debated and accepted in the Chamber. Each region was to have control over police, health, agriculture and forestry, local communications, museums and by implication over the very important matter of tourism.

The general situation had, however, become explosive in the wake of John XXIII and the Second Vatican Council which had ended in 1965. In many parts of the country young Catholics rebelled against the hierarchy but also against the Christian Democrat Party, for they said that both were a parody of what Christianity should be. Spontaneous groups, *Gruppi di impegno politico*, formed to express this dissent, the *Dissenso* of the young. They were encouraged by certain leftist Catholic writers, such as Raniero La Valle who was in consequence forced to resign from the Catholic review *Avvenire d'Italia*, and by Wladimiro Dorigo, the director of *Questitalia*. There was a meeting of the new Dissenters, not surprisingly at Bologna, in February 1968, but also in other places. They expressed tremendous indignation over Vietnam and the fact that the Italian Government condoned American action by remaining in the Atlantic Alliance. The young idealists spoke increasingly of a *Repubblica Conciliare*, a Conciliar Republic, which should turn Pope John's dream into reality by drawing in all men of goodwill from all parties. It was early in 1968 that Cardinal Lercaro of Bologna, who was seventy-six and thus a year older than the Council had decreed for retirement, resigned. Lercaro was almost a 'Dissenter' himself but his resignation made it easier to oblige Ottaviani to withdraw from his position of Secretary of the Inquisition or Holy Office. Ottaviani was succeeded by the Croat, Seper, but the appeasement of the spontaneous groups of dissent was not achieved by the ecclesiastical authorities. It was comic to find mildly sexy stuff introduced even into rural parish magazines, in central Italy at any rate, no doubt in order to placate the young.

The Communists sniffed the air with relish, idealists among

them with gratification, intriguers and dogmatists with interest, for ideals that could be used to their advantage. They worked to revive the Myth of the Resistance, speaking of the need for a new liberation of Italy from the menace of men like De Lorenzo.

The idea of a Catholic–Communist Conciliar Republic filled the older generation with despair; indeed on reflection it seemed to suggest a new version of the corporative state. The *Corriere della Sera* on 11 February devoted a leader to the subject in which it insisted, as the *Corriere* was bound to, upon the necessity of keeping Church and State apart. Six days later the *Corriere* published a letter from the Socialist writer, Silone, which in view of his deeply religious feeling was of unusual interest. He too urged what he called the widening of the Tiber [1] to keep State and Church apart, for he did not believe that the Catholic hierarchy had really changed for the better. Nor did he feel any confidence in the Communists. He spoke of the desperate plight of the Sicilians of Gibellina where an earthquake had occurred in January, but neither Church nor State had yet come seriously to their rescue. Finally he expressed his opinion that the introduction of divorce into Italy was essential; in 1967 a measure in its favour had been introduced without success into the Chamber by a Socialist deputy called Fortuna who was supported by the Liberals. It was clear that Silone had no sympathy but only suspicion, at the best a feeling of irony, for the *Repubblica Conciliare*.

The general election was held in May 1968. The parties on the Right lost, the Christian Democrats and Republicans made slight gains, while the Communists, as well as the Party of Proletarian Unity, made considerable ones; the number of Communist deputies in the Chamber was now up to 177. The fiasco once again was that of the supposedly reunited Socialists. The reunion at the end of 1966 had looked wholly unconvincing and the Social Democrats now lost one and a half million votes by comparison with the voting for Nenni's followers together with the Social Democrats in 1963. This was a crushing blow to the whole *Centro-Sinistra* and the end of Moro's Government. Once again a caretaker Christian Democrat administration, this time led by the hitherto President of the Chamber, Giovanni Leone, was installed in order to keep the country going until the Socialist

[1] The Tiber flowed between the Vatican and the Quirinal.

congress due in October. In fact Leone remained in office until 19 November after the Socialists had made an even worse exhibition of themselves at their congress.

Meanwhile on 21 August 1968 the Russians had invaded and occupied Czechoslovakia. Italy was as greatly shaken as by the events in Poland and Hungary in 1956, perhaps more profoundly. This time there was no parallel Western folly like the Suez invasion to soften the brutality of the Russians' action. What the Czechs had brought about under Dubček was, at any rate in theory, not unlike life under the Communist mayors in Emilia. But in Bohemia the Russians forbade this and were soon to suppress it odiously. The PCI and the CGIL made their condemnation clear, although *Unità* padded its statements with abuse of 'the soiled hands of the Italian bourgeois press' which was now mocking at the Russians. It was characteristic that La Pira sent a telegram of protest to Kosygin, and a leftist Catholic deputy called Corghi, who had ostentatiously resigned from the Christian Democrat Party in support of the *Dissenso* earlier in the year, condemned the Russians roundly. Not even this situation could make the Socialists recover their sense of politics. The clarity of the Italian Communists' response compared favourably with the murkiness of that of the French Communists: it was an expensive decision, as it was likely to mean that noticeably less money would come from Moscow to the offices of the *Partito Comunista Italiano*. Temporarily the Russian occupation of Czechoslovakia strengthened young extremists, either the *Missini* on the Right or the young on the Catholic Left or on the so-called Chinese Left which despised Russia and her admirers as conservative – imperialist too, they might have said, but that word could not be divorced from the Americans or at least their government.

The obvious person for the post of Prime Minister was Emilio Colombo, the brilliant economist who had been a highly successful Minister of the Treasury for years. Moro, however, who was playing for popularity with the young by toying with the idea of the Conciliar Republic and smiling at the Communists, torpedoed Colombo by intrigues discrediting him as reactionary – he was certainly a devout and orthodox Catholic. The net result was that the Christian Democrat Secretary-General was nominated Prime Minister, inaugurating a brief Rumor era in December 1968 which was to last until July 1970. Rumor's first Government was

sworn in on 12 December 1968 and he then went home to celebrate this and Christmas at Vicenza. His Cabinet was a very big
one with twenty-seven ministers whose salaries were increased
sharply. Nenni became Foreign Minister and De Martino Vice-
Premier so that the Socialists with nine ministers were over-
represented, especially De Martino's followers. Colombo was
back at the Treasury but Moro and his friends were excluded.
Rumor himself was soon succeeded as Secretary-General of his
party by Flaminio Piccoli, who, however, did not last long. The
old Socialist stalwart, Pertini, was chosen as President of the
Chamber and Fanfani as President of the Senate. Oronzo Reale,
the only Republican to remain in the Government, was Minister
of Finance. His party chief, La Malfa, expressed considerable
disapproval of the Rumor Government in spite of the fact that it
represented a reaction against Moro and the dialogue with the
Communists aimed at bringing them into the Government.

By now, incidentally, Moro and the Dialogue were supported
by Livio Labor, the President of the ACLI which had once
demurely followed the behests of Pius XII. Earlier in the year
at Vallombrosa it expressed its change of front; for ACLI also
advocated the Conciliar Republic, and its members were freed
from their old allegiance to the Christian Democrat Party.

From its beginning the Rumor Government was unsatisfactory.
Rumor himself, ever aware of the possibility of a new dictatorship, scarcely concealed his anxiety as the divisions within both
the Christian Democrat and the Socialist parties became ever
more conspicuous. On 26 January 1969 the Prime Minister
warned protesting youth that their fellows behind the Iron
Curtain suffered far greater wrongs – it was not long since the
Czechoslovak student Jan Palach had burnt himself to death in
a supreme Hussite gesture. Other ministers expressed their
sympathy with the grievances of Italian students whose *contestazione* or challenge to society continued in the shape of much
demonstration on every possible occasion. As in other countries
the lesser injuries to freedom perpetrated in the United States
excited furious indignation, while those for which Russia was
responsible throughout Eastern Europe were scarcely noticed by
the young people of Italy. Educational reform was discussed in
the Chambers at great length. This also exacerbated the dispute
about co-operation with the Communists who had put forward

plans for student participation; Riccardo Lombardi, for one, favoured co-operation with them and so did some leftist Catholics, although no one at that time holding ministerial responsibility. The question of university reform severely shook Italian society. Typically the Communist Party encouraged extremist agitation except where it was locally responsible; the Communist Mayor of Bologna, Fanti, the successor of Dozza, in his turn condemned demonstrations by Maoist students. Violence was inaugurated by the explosion of a bomb at the Ministry of Justice on 31 March 1969. In April serious riots broke out at Battipaglia near Salerno in the south when a tobacco factory was closed down. This was followed by outbreaks of violence all over the country during which the police showed themselves too ready to shoot. The Communists denounced every mistake made by the police as part of a Fascist plot organised by men like De Lorenzo. In the middle of April, just when Colombo had said that there could be no question of bringing the Communists into Rumor's Government, a junior Christian Democrat Minister, an Under-Secretary at the Ministry of the Interior of all places, spoke at Florence in favour of including them. This was De Mita of the *Base corrente*, a friend of the leftist journalist and politician Donat Cattin, and of the former Minister of Instruction, Sullo, who had resigned in March. To Rumor's dismay, his deputy in the Cabinet, De Martino, came out at the end of April in favour of bringing the Communists into the Government, and was echoed by a former Social Democrat, now a member of the PSI, the Calabrian Mancini. This heralded the collapse of Socialist reunification after all at the beginning of July. Nenni was obliged to acknowledge his final failure and withdrew to his country home at Formia: he was now seventy-eight but was considered to have done unusually well at the Farnesina since Christmas. Two of his Social Democrat colleagues, Tanassi and Ferri, much abused by the Communist press, resigned to found the *Partito Socialista Unitario* (PSU), welcomed by Malagodi as the rebirth of democratic Socialism; the PSU founded a newspaper called *Umanità*. Rumor himself had resigned on 5 July. 'The situation is chaotic', said the *Corriere della Sera* on 6 July; earlier it had written: 'There is a limit to everything, even to the cannibalism of the parties.'

Rumor had certainly done his utmost according to his lights, preserving a modicum of public order, giving pensions to the neediest old people and working on endless reform projects. His life had not been made easier by the Socialists again putting forward in May their proposal to introduce divorce. As a *dévot* he accepted the claim of the Roman Catholic Church that divorce was unconstitutional, though he can scarcely have welcomed a Vatican dictum likening divorce to the 'mass production of pornography'. Rumor had worked very hard, but as *The Times* correspondent in Rome wrote on 7 July: 'To the ordinary Italian, however, politics are increasingly becoming an excuse for not governing.'

The Communists, of course, claimed that the fall of Rumor's Government was 'all a Fascist plot'; men like De Lorenzo, they said, were clearly getting ready. By furious attacks upon those like Tanassi who were struggling to keep the Communists out of power they seemed to wish to paralyse the Republic; they accused Tanassi of having been bought by Nixon. At the same time it was evident that they did not wish to accept any of the responsibility of government except locally in central Italy. Saragat asked Rumor to form a new government; the Communists hoped it would include leftist Christian Democrats and the pro-Communist followers of De Martino who would wish some posts to be offered to the Communists. Both Saragat and Rumor were determined to avoid this and, after a month of negotiation, on 5 August Rumor was able to announce that his second government would consist of Christian Democrats only for whom the former *Centro-Sinistra* parties had promised to vote. Moro succeeded Nenni at the Farnesina, and the leftist *correnti* of Christian Democracy were much more strongly represented than before. It was reckoned that at least seven groupings of the *Democrazia Cristiana* were present in the new government. Perhaps the most interesting new appointment was that of Carlo Donat Cattin as Minister of Labour; he belonged to a leftist *corrente* called *Forze nuove* which published a journal of that name. As Minister of Foreign Trade a man of only just thirty-seven, Riccardo Misasi, was appointed; this was a record in youthfulness among Italian ministers, although Misasi belonged to an older *corrente*, the *Sinistra di Base*. These appointments did not solve the problem of how to deal with the Communists, for the young leftist Catholics were still susceptible to the formula of the Conciliar Republic. Rumor created a bad impression by naming fifty-five under-secretaries. Since he was certainly an upright man who was not out to sell favours, this must be taken to indicate in advance the weakness he would reveal in 1970.

Rumor presented his new Government to the Chambers on 8 August 1969: he said that he planned to revise the Concordat in connection with divorce, and to prevent the new outflow of capital. He made it plain that he regarded his new Cabinet as something preparatory to a restoration of a multi-party Centre-Left government, and not only *La Stampa* but also the *Corriere della Sera* commented that this was indeed the only practical possibility. In both Chambers both Socialist parties supported Rumor and he won a comfortable majority. However, all the commentators hastened to point out that the Communist problem in Italy was no nearer a solution. As De Martino said in the Chamber, it was impossible to isolate so large a block of voters even if their leaders were equivocal in their attitude towards Russia – by now condemnation of the Russian occupation of Czechoslovakia was more faintly expressed by the leaders of the PCI.

The Italian economy was surprisingly buoyant, but in spite of the great efforts that had been made to stimulate southern economic life, the north and centre were still advancing twice as fast as the south. At the same time southern immigration was still creating an acute housing shortage in the so-called industrial triangle of Milan–Turin–Genoa.

The immediate political issue in August 1969 was that of local elections. Communes, provinces and the new regions were to elect their representatives that autumn, but the financial arrangements for the regions were not yet completed. The parties on the Right, and now Tanassi's Social Democrats (now the PSU), at first seemed glad enough for the inauguration of regional government to be put off until the Greek Kalends, but it has been seen that most people with leftist sympathies now hoped for a great deal from the regions, and the PSU in fact ceased to oppose them. The PSU deputy Matteo Matteotti, son of Mussolini's famous victim, reminded the Chamber early in September 1969 that the Republicans had tried to insist upon the abolition of provincial councils before creating regional ones – it was obviously foolish to have both.

When the new young Communist leader from Sardinia, Enrico

Berlinguer, answered Rumor in the Chamber on 9 August, while rejecting the notion of the Conciliar Republic, he was not intransigent about affairs at home; it was over foreign policy that he demanded an abrupt change – Italy must become neutral as between the U.S.A. and the U.S.S.R.

On 21 August, the very anniversary of the Russian invasion of Czechoslovakia, the 'moderate' Communist leader Giorgio Amendola, speaking in the Chamber, demanded that the Communists should be included in the Government. On 30 August Piccoli, Secretary-General of the Christian Democrats, rejected this, as he called it, 'arrogant request' which would place Italy at the least in the orbit of Moscow. Piccoli appealed for a new launching of the Centre-Left as something 'organic' – in this period much was said about an organic Centre-Left government, though it would be difficult to feel certain what was meant by this constantly used phrase. In an interview with La Malfa, more and more the elder statesman of the day, which was published in the *Corriere della Sera* on 19 September, he was quoted as also advocating this *rilancio* or relaunching. He complained, too, that the Communist Party needed to revise their slogans, bringing them into line with conditions in a developed Western country, as most of Italy, even patches in the south round a big new factory in Brindisi or Taranto, had become.

Actually the Communists had their own fresh opening in September 1969. Many labour agreements made in 1966 terminated now after three years. Rumor had been warned to expect a 'hot' autumn and now it came, with millions of people on strike. The Italian trade unions were too poor to distribute regular strike pay; hence most strikes were for short periods, often for twenty-four hours or less. In some ways the organisational difficulties of industry were almost as great as, though of course different from, those in Britain during longer strikes. Rumor made a few quite sensible speeches about higher wages requiring greater productivity; he was also ready, he said, to consider greater participation by the workers. Yet most of the autumn passed with ministerial speeches only about the relations of the parties. The PSU quarrelled with the PSI; within the Christian Democrat Party the squabbles between the *correnti* were unending. Nenni re-emerged once from his country home at the meeting of the PSI Central Committee on 8 October when he emphatically, almost desperately, agreed with La Malfa that the *Centro-Sinistra* should be relaunched as soon as possible. Three weeks later he spoke in the same vein at Milan, alluding more clearly to the terrifying similarity of the present to the pre-Fascist period early in the twenties. Rumor was all too conscious of this similarity, which seemed indeed to wear him down.

By this time Colombo had made an important speech at Turin on 13 October with which he hoped to restore national confidence and in which he insisted once again on the importance of planning. A few days later, however, Colombo took what appeared to be a very imprudent step, for he quarrelled quite openly with the Prime Minister in withdrawing his support from the Christian Democrat Secretary-General, Piccoli; the latter was being attacked as weak and inflexible by the leftists in the party, particularly by the group of *Forze Nuove* led by Donat Cattin. With so many more urgent tasks to perform, the governing party held a national council on 6 November at which Piccoli was replaced as Secretary-General by Arnoldo Forlani, a protégé of Fanfani. The change was, however, partly sponsored by Fanfani's major rival, Moro, who increasingly foregathered with Donat Cattin and his friends. Forlani could only say, as all but the PSI and the Communists and PSIUP did, that there must be a return to the Left-Centre coalition. Donat Cattin was not altogether pleased to hear this as he favoured a Christian Democrat–PSI two-party Government; this was in fact bound to depend upon the Communists, who began to appear irresistible.

For the strikers – the most important were the metal-workers – had been out on strike on and off since the beginning of September. The Communist leaders amused themselves with abusing the 'intransigent' refusal of the employers to offer terms, a refusal which, according to *Unità*, could only be intended 'to provoke the workers' and ruin Italy. Not only were the working people making their traditional claim to 'participation'; when they announced the general strike which they held on 19 November 1969 they made a quite new demand: they said this strike was intended, not to influence wages and conditions of work, but rather to overcome the Government's incompetence and compel it to put through long-talked-of housing measures. This was to be a foretaste of the spring of 1970. It brought a triumphant note into the declarations of the Communists who held their Central

Committee meeting at the end of November. In an article in *Rinascità* Longo wrote: 'There can be no discussion which does not take its lead from the issues posed by the people, the workers; we must unite to work out a concrete policy of renewal and progress based on a new majority of the left.' The words used by the Communists varied very little. There seemed to be only one fly in the Communists' ointment, the splitting off from them to the left of a pro-Chinese group led by Pintor, Natoli and a woman called Rossanda: the group was named after its publication, *Manifesto*. It claimed that the PCI had become part of the Italian establishment, which was reactionary: Berlinguer replied by condemning the *Manifesto* people as 'adventurers'.

While the strikes continued, young people of various extremist persuasions carried out violent demonstrations; the police hit back at strikers and students, and a southern policeman called Annarumma was killed in Milan. On 9 December there was a debate in the Chamber on public order, the Minister of the Interior, Restivo, pointing out that Togliatti, too, had always condemned hooliganism (*teppismo*). Three days later two bombs exploded in the Agricultural Bank in the Piazza Fontana, Milan, killing fourteen people; there was also an explosion in Rome at the monument to the Unknown Soldier, but only two casualties there. The shock of the Piazza Fontana incident was very great; the police arrested about 150 people in Milan and about half that number in Rome, though most of them were soon released. Libertini of the PSIUP declared the explosions were a new Reichstag Fire, though he can have had little idea of what he was saying. On the whole, as the *Corriere della Sera* wrote, Italy stood the test. It would have been a wonderful opportunity for a *coup d'état* and was indeed said to have been created by the Greek colonels. Once again the PCI took good care not to take over governmental responsibility. Rumor, returning from the funeral of the fourteen in Milan on 15 December, met with the secretaries of the four Centre-Left parties on the same afternoon: he begged them to join up again so that a stronger government could adopt a planning programme. Indiscreet correspondents representing the *Observer* and *Die Zeit* had published in those papers the Communist indictment of Saragat according to which his intrigues had caused the Socialist division and fortified the resistance of the employers to strikers' demands. When on 18

December the President of the Republic held his great annual reception to celebrate Christmas and the New Year, his stock on the contrary stood high. It was significant that at the head of the list of his guests, among the names of the foremost magistrates, the *Corriere* reported the presence of the presidents of the regions and of the regional assemblies: an important reform had in fact been launched with little other publicity. The year had begun with Rumor's first Government; it was ending not unhopefully with a certain new confidence. On 21 December one million privately employed and 300,000 employed by the state ended their strike with an agreement which provided higher wages and shorter hours.

However, during the early months of 1970 everything gained seemed again to be lost. By the end of January Rumor resigned – it was said that he was nervously exhausted, but it was also whispered that Fanfani and Moro and even Forlani had ceased to support him. The leftists of the *Democrazia Cristiana*, Donat Cattin and others such as Galloni, had continued to make fairly wild suggestions about co-operation with the Communists, and Lombardi was said to have seen Berlinguer in private. The worst difficulty now seemed to be that the lay parties were pressing for divorce to be made legal while the Pope early in February issued an uncompromising veto saying again that this would infringe the Concordat. This question made things even more difficult for the extremely Catholic Rumor and seems also to have precipitated his resignation. Giulio Andreotti, long a minor leader among the Christian Democrats and leader of their Deputies in the Chamber, won himself some prominence during February 1970 by seeming to show that negotiation with the Vatican on this issue was possible. Pope Paul VI, as so often, vacillated. Meanwhile, in spite of the settlements with the *metal-meccanici*, strikes and demonstrations were starting up again, not surprisingly since the politicians had really discredited themselves. Increasingly the trade unions and the students declared that if all reforming legislation was blocked by politicians' intrigues they must push it through by their strikes and demonstrations.

Early in February 1970 Saragat had instructed Rumor himself to try to re-form a Centre-Left four-party government, but the question of divorce seemed to have poisoned the air, though in fact the bill in its favour drafted by the PSI Socialist, Fortuna,

and the Liberal, Baslini, a year or so earlier, was accepted by the
Chamber. As the *Corriere della Sera* wrote on 15 February, the
issue of divorce had become confused with that of the revision of
the Concordat. Wistfully it recalled that John XXIII had
preferred to regard Italy as external to the Church, simply as a
country with a big proportion of Catholics. When the Liberal
leader, Malagodi, spoke a month later of the 'monstrous
irrelevance of divorce' to Italy's political crisis, opinion seemed
to be as much on his side as on that of the Vatican.

At the end of February Rumor seemed to give up his new
attempt, and during March 1970 the President charged first
Moro and then Fanfani to attempt the reconstruction of a Left-
Centre Government. Both were obliged to admit inability to do
so. Fanfani's behaviour was as usual a mixture of sinister and
absurd. It was general knowledge that he intended to succeed
Saragat as President and was therefore more inclined to posture
now than to try to govern the country. Apart from the fact that
he seemed increasingly to advertise himself as a Man of Order,
he estranged sympathies by trying to insist that the party secretaries should all four become ministers *ex-officio:* it may be
remembered that he himself had once combined the position of
Prime Minister with that of Foreign Minister and Party Secretary.
This inclusion of the party secretaries would have meant an even
bigger Cabinet and at that time recommended itself to no one else.
Although most politicians were urging the addition of regional to
the communal and provincial elections which were overdue, no
one – not even the Communists – wanted a general election. As
Pertini, the splendid old Socialist who presided over the Chamber,
said, the Parliament elected in May 1968 had been a good one,
and the risks in holding another election before it was due in
1973 were great, as would be the expense. By 18 March Fanfani
was obliged to abandon any pretence of forming a government
and the unfortunate Rumor was asked on 20 March to try again.
In desperation he succeeded within a few days and by the end of
the month the new *quadripartito* Government was in office. Moro
went back to the Farnesina, and the Republican, Oronzo Reale,
became Minister of Mercy and Justice again; these two positions
gained in importance because the two ministers were to negotiate
with the Pope about the Concordat and the Divorce Bill.

In the statement of his policy which Rumor made on 7 April

he revealed that the Pope had already protested against the introduction of divorce into Italy in 1966 when Fortuna first mooted
it, and again in 1967. It was, in fact, on 30 January 1970, a week
before this was made public, that Paul VI had condemned it as
a breach of Article 34 of the Concordat. Now Rumor stated that,
after Moro and Reale had discussed the question with the Vatican,
Parliament would have the last word. Finally a referendum
might be held.

This speech of Rumor's was Rumor at his best. Speaking of the
amnesty which the Left had been demanding for strikers' offences
because the police were felt to have been too severe and indeed
Mussolinian, he declared that there would certainly be an
amnesty: it would apply to all those who had not committed
crimes which offended the conscience of society. At the same
time the punishment of drug offences, which had become a Mafia
activity, would be made harsher. In foreign policy the promotion
of British membership of the Common Market would be intensified and steps would be taken towards Italian recognition of Communist China: he did not refer to the revolution in Libya in the
previous September which affected Italy very disagreeably. An
economic programme with enough planning in it to please the
Communists was put forward, and plans for educational reform.
A 'Labour Charter' guaranteeing all essential workers' rights
was in fact carried through on 14 May 1970. It incorporated
earlier legislation but still did not allow management by the
workers.

It was, however, about the regions that Rumor, not unexpec-
tedly, spoke with greatest warmth; it was only in connection with
the regions that there were interruptions in the Chamber. 'The
regions,' Rumor said, 'are charged with hope for us although
they represent uncertainties . . . within two years the government
will transfer to them their administrative functions', having by
then worked out the administrative details. It was a progressive
speech in the truest sense and people on the Right felt that the
Conciliar Republic was indeed approaching. There was a substantial vote of confidence in both Chamber (17 April) and Senate
(10 April).

Two difficulties, familiar enough, presented themselves
immediately. As the *Corriere della Sera* pointed out on 19 April,
no one had paid serious attention to La Malfa's warnings that

the provinces should be abolished when the regions were set up. In consequence the regional structure was now superimposed upon the provinces, and prefects or their colleagues tended to appear as regional officials: if this went far the regional reform would become meaningless since the prefects were nominated by the Ministry of the Interior, not elected.

Almost worse than this was the fact that the first task of the newly reunited Centre-Left Government was to hold, together with communal and provincial ones, elections to the regional assemblies on 7 and 8 June 1970. These elections, postponed since the autumn, were a contest between the coalition parties as well as between them and the other parties of whom only the Communists had importance. And the most acute issue between the coalition parties was that of the *giunte* – i.e. were the coalition parties free to ally themselves with other parties in local governments? Already in the second half of April 1970 the *giunta* or municipal council of Ravenna caused an outcry. Ravenna was something unusual, a stronghold, virtually the only one, of the Republican Party, for there were fourteen Republican members of the town *giunta* to eight Christian Democrats, and the mayor had been Republican as long as anyone could remember. But now he resigned, and the PSI, De Martino's Socialists (no longer Nenni's), left their alliance with the Republicans and Christian Democrats and entered into one with the Communists and PSIUP. The moderate Christian Democrats, such as the Prime Minister, agreed with the Republicans and Social Democrats in condemning the PSI for letting the Communists in, whereas Christian Democrat leftists like Donat Cattin felt that he or the Socialists should be ready to meet the Communists half-way, partly in order to restrain Communist influence. This quarrel inevitably dominated the weeks leading through May to the June elections. Already Fanti, the Communist Mayor of Bologna, had announced that the PSI had agreed with the PCI and PSIUP to participate in the government of the region of Emilia. The Communists could hit back, as they of course did, by referring to the tolerance by Christian Democrats and Social Democrats of a neo-Fascist member of the town council of Pavia. It was, however, customary in Italy to let town councils be a proportionate reflection of party strength, and not, as they would be in Britain, a clear-cut question of Government and Opposition. (It should perhaps be noted that trade union officials were ineligible for election to local governments.)

Meanwhile strikes continued, ostensibly in favour of reforms. In May there was a brief civil servants' strike. On 23 May the Vatican revived the divorce controversy by launching a series of broadcasts against divorce. This was fiercely attacked from the Socialist side as in itself a violation of the Concordat.

As the elections of June approached, the Christian Democrats became more anti-Communist, and Longo on 5 June abused their 'thirst for power'. On 24 April the *Osservatore della Domenica* had condemned the idea of a Conciliar Republic. Another event to be noted was that at the beginning of June Donat Cattin reversed his attitude and began to condemn the incessant strikes: he was an arch-opportunist and this probably indicated that the public was becoming exasperated.

On Sunday 7 June and the morning of Monday 8 June elections were held for most of the town and provincial councils and for fifteen regional assemblies – the other special regions, like the islands and Aosta, were already provided for. Although many voters had three voting papers and the rest two or one,[1] 91·3 per cent of the voters voted. After the confusion and the turmoil of the last two months they voted prudently. By comparison with the general election of May 1968 extremism seemed to lose. Thus the PSIUP whom the Communists had cold-shouldered lost, though not the *Missini* who gained the slightly alarming percentage of 5·2 (in 1968, 4·3) in all. The Christian Democrats lost a very little, the Communists almost nothing, though they did lose and not gain. Both kinds of Socialists and the Republicans gained so that the lay parties of the Government coalition were fortified – it was a vote in favour of divorce. Leftist extremists of the *Manifesto* kind did not stand for election. The official Communists were as strong as ever in their Red Belt where the moderates like Amendola prevailed. Here they had ruled locally for years and would now do the same at the higher regional level. In actual fact the regional assembly of Emilia had a leftist, i.e. Communist plus PSIUP, majority of one, so that it was independent, more or less, of the behaviour of the PSI. It was in

[1] In Rome, Foggia and Ravenna the electors only had voting papers for the new regional assemblies, other local elections having taken place recently.

Tuscany and in Umbria that the Communists plus PSIUP and the Centre-Left were exactly balanced and the situations precarious; it was more particularly so in Umbria which had suffered more than the rest of central Italy from desertion of hill-farms, so that it was in fact sinking down towards *Mezzogiorno* levels without receiving help from the *Cassa per il Mezzogiorno*. Communist local government had not been able to prevent this; it had failed to establish sufficient new industries in Umbria. One other election result of June 1970 must be noted. Sesto San Giovanni, a working-class suburb of Milan, was a symbol of Communist strength. On this occasion, however, the Communists lost two seats in the *giunta* for the town and the *Centro-Sinistra* gained proportionately.

On 6 July 1970 Mariano Rumor suddenly resigned again. It was said that he had done so without consulting his colleagues. Italy gasped; capital was rushed out of the country once more. It is generally supposed that the nervous strain imposed upon him ever since he first became Prime Minister in December 1968 had been too much even for his equable temperament. Some said he hoped to force a clarification; others remembered that he had often said he kept his things packed ready to retire to Vicenza at a moment's notice. It was known that he had found the behaviour of the PSI exasperating, particularly since the party (except of course for Riccardo Lombardi) had fought the local elections in the name of the *Centro-Sinistra* and had then not stuck to it.

At least the trade unions called off the general strike they had planned for 7 July, which had in any case lacked the approval of the UIL, the Social Democrat trade union organisation. There followed a new crisis, lasting a month this time, a crisis which was only of interest for one or two reasons. This is perhaps the point at which to say that although Italian Governments seemed to change so often, the actual minister in any one ministry seldom changed, so that there was more continuity than was sometimes supposed. And the last Cabinet carried on until the next one was formed, without perhaps sufficient conviction.

After Rumor resigned on 6 July the President of the Republic at first entrusted Andreotti with the task of forming a new four-party government; when after a fortnight Andreotti was obliged to admit failure, a few days later Saragat asked Colombo to try to form another Centre-Left government. Saragat, himself a Social Democrat, was often under fire, especially from the Communists, for interfering more in politics than the President should. It was said that he should in no way define the task of the man to whom he offered the Premiership, but this could not be constitutionally established. Moreover Saragat made his choice and defined his terms only after long consultations with all shades of opinion.

In the background all the time there lurked Fanfani's determination to succeed Saragat. Since the President was elected by both Houses of Parliament in joint session a good deal depended upon whether Fanfani believed that the members of the existing Chambers were more likely to elect him than a new Parliament. If Saragat felt, or could be persuaded, that the unceasing crises called for a dissolution, the constitution required him to dissolve not less than six months before his own Presidency ended, i.e. not later than June 1971. There was obviously a good deal of consultation between Saragat and Fanfani in July 1970. It was significant, also, that Fanfani was known to be moving to the Right, expressing views in favour of curbing the trade unions. If he increasingly appeared as a man of order, his rival, Moro, Foreign Minister under Rumor, kept carefully to foreign affairs, but made it evident that the Left could turn to him as future President.

Andreotti, formerly one of De Gasperi's young men, indeed one of the youngest of them, and often before this Minister of Defence, had moved from the Right of the *Democrazia Cristiana* more to its Left. Indeed he now enjoyed some mysterious favour with the PSI which at this time was extending its alliances with the Communists, for instance in the city of Prato near Florence and in the provinces of Florence and Perugia. Yet, in spite of his personal link with Rome – he was the only Roman among eligible Cabinet ministers – his attempt to form a government in July 1970 all the time had an air of unreality.

Colombo was another matter. He was just as young and had very nearly had the chance of forming a government less than two years earlier. He was a southerner from Lucania with a well-deserved reputation for integrity and intelligence. He had now been the economic specialist in the Government for years and a good friend of Carli, the able Governor of the Bank of Italy.

Colombo had been labelled reactionary, but during the strikes of the autumn of 1969 he had said that the nation could afford higher wages. Colombo was a devout Catholic, yet certain that there should be a clear line of division between Church and State. The Communists all this time kept up their usual cry of 'Danger from the Right', reminding the public of 1960 which was not comparable. By 22 July it was evident that Andreotti would not succeed in forming a government, and a couple of days later, after the customary consultations, Saragat entrusted Colombo with slightly more room to manoeuvre. By 5 August he succeeded in inducing representatives of the four Centre-Left parties to serve under him. Colombo's biggest disappointment was that La Malfa refused to succeed him, Colombo, as Minister of the Treasury. By and large the new Government was the same as Rumor's with Ferrari Agradi at the Treasury and a place found for Matteo Matteotti of the PSU as Minister of Tourism and Entertainment. This time, alas, there were fifty-eight under-secretaries. In his first official statement on 10 August Colombo insisted that his Government rested upon principles with which those of the PCI could not be reconciled. On the following day he indicated that he would shortly put through an emergency economic decree. It was a very hot summer in Rome and difficult, therefore, to keep the politicians at work in August. Colombo, however, persisted, and on 27 August he published his emergency decree which aimed at reducing the deficits being incurred by various public *Enti* and at increasing incentives to production. Before everything, taxes on what were considered luxuries were increased, for instance on petrol, telephones, passports, bananas, spirits, luxury houses and motor-boats. Everywhere the Left retorted with factious posters saying *No al Decretone* which had immediately come into force but required subsequent confirmation from the Chambers.

Colombo moved quickly on to the most necessary reforms and by the beginning of September ministers were at work on health, housing and transport. On 10 September he opened the annual Trade Fair at Bari with appeals for social peace in order to carry out Italy's new plan which must aim before everything at integrating southern Italy into the Italian economy: he insisted that the problem of the south was the problem of Italy herself. Until it was solved nothing would really make sense. As Colombo spoke his audience felt that here was the first Prime Minister from the deep south – Moro had come from the relative comfort of Bari. And Colombo was single-minded, technically competent and clearly of great energy; perhaps he could solve Italy's problem of the *Mezzogiorno*.

In the midst of unsolved problems and the continuing growth of a lay spirit induced by economic development, Italy approached the centenary of her capture of papal Rome on 20 September 1870. Saragat read a long speech to all the notables of the nation gathered at Montecitorio, in which he praised the *Risorgimento* and the Resistance. Everyone agreed that the Church had been given a new and better life by losing its temporal power, and Paul VI defied the tradition of Pio Nono by the well-worn exhortation to render unto Caesar the things that are Caesar's and to God the rest. In the *Corriere della Sera* famous scholars made their contributions to the occasion. A. C. Jemolo, distinguished historian of the relations of Church and State, had boasted in the past that although he was a devout Catholic he had never voted for the *Democrazia Cristiana*. Now he declared in the *Corriere* on 20 September 1970 that both Church and State were collapsing, unable to help each other; he spoke with scorn of priests who married as forgetful of Christ's words 'My kingdom is not of this world'. In the same paper on the same day – how characteristically Italian to criticise themselves! – someone else wrote that '*Roma capitale*', the cry of 1870, was a hundred years later a capital city without a real road structure, without the underground so often proposed, without the new universities it needed, with half a million people living in disgraceful dwellings and 50,000 in barracks. The Rome of Pius IX had been a slovenly village by comparison, but the Rome of Saragat and Paul VI was the uncomfortable home of an exploding society, erupting vigorously through the crust of politicians' words. At the beginning of December divorce was accepted by both Chambers and the lay aspect of the nation established at last. In the last hundred years the wheel had turned full circle.

A last-minute attempt to put back the clock was made just after this, almost a serious Fascist *coup*. It was organised to take place in the night of 7–8 December 1970 by a notorious naval commander who had played a prominent part in Mussolini's

'Social Republic' or Republic of Salò, dominated by Hitler from September 1943 to April 1945. In those days Prince Valerio Borghese was still in his thirties; he had been a brilliant submarine commander and was still in charge of a naval formation called the *Decima MAS*. By 1970 he was sixty-four and the head of a National Front. On the night of 7 December he called his men together in Rome and was waiting to emerge from his flat near the Termini station to lead them; he had planned to capture Restivo and Vicari, the chief of the police, when some politician of weight persuaded him to desist because the feelings of the Army were against him, as were those of the trade unions and the big majority of the politicians. After two hours of waiting in vain, some of his followers consoled themselves with a big *spaghettata* or spaghetti meal. Almirante, the leader of the MSI, did his best to dissociate himself and his followers from Prince Borghese.

9 Further Development of the Italian Economy

The economic miracle in Italy collapsed in 1963. Briefly the causes for this were the substantial wage increases in 1962; these were followed in 1963 by an increase in imports of 25 per cent, consumer prices rising by 8 per cent. The development of the home market, which had earlier been a source of strength, now diverted goods from the export trade. The crisis of confidence among the industrialists was accentuated by the actions and plans of the new Left-Centre Government, by what Nenni, who was second-in-command in Moro's first administration, called *programmazione*. It has been seen that this government took office in December 1963 and large sums of money were promptly exported by hostile financiers and businessmen. Thus in 1964 Moro and his colleagues found themselves with the need to restrict credit instead of investing generously in planned reforms. On 23 February a ministerial decree cut down hire-purchase, raised the price of petrol and increased the taxes on larger cars and motor-yachts. On 14 March an American loan of $1,225 million, which included credits from the International Monetary Fund, was announced. At about this time Innocenti and other car producers, including even Fiat, began to reduce the hours worked in their factories; so did Magnodyne, a firm which made television sets. Olivetti had run into difficulties earlier. ENI, Pirelli, the makers of optical instruments, Montecatini and Edison (which had taken to chemicals since the nationalisation of electricity) continued to do quite well. In spite of a pessimistic letter from Colombo to Moro which was published in the Roman paper *Il Messagero* in May, by June foreign capital was flowing back into Italy and exports rose to a record value of 319·6 milliard lire. By the end of August gold reserves had risen and there was a favourable balance of payments of about 150 milliard lire which had risen to 184 milliard by the end of September 1964; there had been a deficit of 500 milliard at the end of September 1963.

Interesting figures were published by the Italian car industry in the middle of September 1964. In the first seven months of the year 216,361 motor vehicles of different kinds had been exported, 11·9 per cent more than in the same period in 1963. By now the cost of Italian industrial production, according to the President of the Bank of Italy, had been pushed up by a 43 per cent increase in wages in 1962 and 1963 taken together – wages had thus increased more than twice as much as prices. Moro had therefore decided just before this that the State should take over a substantial proportion of the employers' contributions to their workers' social insurance; in this way he gave the industrialists a subsidy worth about 200 milliard lire.

By 1965 expansion was continuing, and a Five-Year Plan was supposed to come into operation at the beginning of 1966. Although the gross national product had only increased by 3 per cent in 1964 – the increase was more like 1 per cent in 1963 – the Plan was based upon a 5 per cent annual increase and proved unnecessarily cautious, for production was now to soar to fresh heights. There was also an incorporated Green Plan (*Piano Verde*), the second one, for agriculture.

Since the Edison electrical company had gone chemical, it combined in 1966 with the great chemical firm of Montecatini into the 'Montedison', and two years later in 1968 ENI, through buying shares, became a determining influence over and within Montedison, which thus became a monster chemical concern based on ENI's gas and oil. By 1969 the products of Italy's chemical and petro-chemical industry reached a value of about 3,000 milliard lire, about six times as much as in 1950. The production of drugs for chemists' shops and of cosmetics added to this figure. This great concern by 1969 contributed 14 per cent of the value of Italy's whole industrial turnover and was by then one of the ten great industrial undertakings of Europe. It was not, of course, an industry offering much employment. Already in 1950 Montecatini had employed about 150,000 people, and this figure had less than doubled for the whole Montedison combine by 1969, reaching only about 275,000.

The southern centre of Montedison was at Brindisi, and ENI

and Montedison, together with the artificial fibre concern SNIA–VISCOSA, had established several factories in Sardinia to produce chemical fibres. The co-operation of these mammoth concerns has made bigger investment, and therefore greater production, possible. Nor has it excluded the activities of smaller firms. One called *Rumianca* and another the *Società Italia Resine* (SIR) established petro-chemical factories in Sardinia in the early sixties, and they have developed satisfactorily since then. The Italian genius is for improvisation, and planning is really alien to it, but in the chemical industry planning has probably succeeded better than anywhere else.

After Mattei's death late in 1962, in itself seeming to many to symbolise the end of the 'miracle', ENI was officially directed by Professor Boldrini until 1966. But the new moving spirit was that of an old partisan comrade of Mattei's called Eugenio Cefis who was only twenty-five when the war ended. Although, unlike Mattei, he disliked publicity, his importance was acknowledged when in June 1967 he was appointed to be the new president of ENI. Apart from the competition of the international oil companies, the 'Seven Sisters' whom Mattei had fought but had been obliged partially to conciliate before he died, Cefis found himself faced with other difficulties. The production of gas in the Po valley, as also of oil in Sicily, was declining, but he arranged to import liquefied gas from Libya (then still friendly to Italy), and gas, as well as crude oil, was soon to be imported from Russia. The Six Days War impeded the import of oil from Egypt and through the Suez Canal in 1967 just when Cefis became President. Then the civil war in Nigeria prevented the production of two million tons of crude oil in the four AGIP fields there for a couple of years. It was in Tunisia that Cefis was able to superintend the production of thirty million tons per annum of crude oil without obstacles. In the circumstances ENI was astonishingly successful, beating all records in 1969. Its turnover grew from 350 milliard lire in 1960 to 1,500 milliard in 1969. After the *coup de main* by which he more or less annexed Montedison, Cefis had become one of the most powerful men in Italy. Shell and other competitors were quite glad to co-operate with him on gas ventures both in the Adriatic and in the North Sea.

The four years following 1965 comprised a period of great

overlapping concentrations in other sectors of industry, not only in Italy, but perhaps with a special intensity there. The process was stimulated by IRI, ENI and ENEL through which, as the *Centro-Sinistra* Government wished, the influence and power of the state increased. Clearly the functions of ENI and ENEL were closely linked since ENI provided most of the raw materials from which electricity was generated, and, of course, Montedison needed large amounts of power. Thus the Italian economy was increasingly both nationalised and rationalised, though both quite incompletely. IRI expanded all the time in many directions, acquiring control of some confectionery firms and of other food industries, Alemagna, Motta and so on: the Cirio fruit and vege-table group was included in these IRI acquisitions. They were made partly to keep out Unilever and some American firms which had big interests here, and partly in order to develop food industries in southern Italy.

During the first five years of its existence, i.e. up to the end of 1967, ENEL had continually expanded in producing electricity and revenue. Its course had not run smooth, for the floods of 1966 had interfered gravely with its work. In 1967 the introduc-tion of summer-time reduced the consumption of electricity; in the same year it has been seen that the Six Days War blocked imports of crude oil, most of which was needed for electricity. Each year more consumers of electricity were connected up, 200,689 in 1966 and 215,712 in 1967, in spite of the obstacles. In 1967, 170 million lire were spent on this. The staff of ENEL had increased from 86,796 in 1966 to 94,277 in 1967, but was likely to decrease in the future. Compensation to former electric companies like Edison had been paid up to date with the help of loans. It should be mentioned that the President of ENEL was Di Cagno, a former Mayor of Bari and a friend of Moro.

In spite of ENI and the gas and oil resources now available, when it came to a fresh wave of development based on planning, the Italian nuclear physicists, proud of Enrico Fermi's contribu-tion to their subject,[1] urged upon the Government the importance of nuclear energy for Italy. A national committee

[1] Enrico Fermi was a highly distinguished Jewish physicist who was driven away from the University of Rome to the United States when Mussolini introduced his anti-Semitic legislation in 1938. Fermi built the first atomic pile.

for nuclear research had existed since 1955: out of this there was formed the *Comitato Nazionale per l'Energia Nucleare* or CNEN in August 1960. When ENEL was established not long after, it became responsible for the country's three existing nuclear power stations. These had been built by Fiat, Montecatini and SNIA–VISCOSA with some American support, at Latina, Garigliano and Trino Vercelles, two in the south and one in the north this time. However, people like Saragat and his friends considered nuclear power stations as too great a luxury for a relatively poor country like Italy – Saragat wanted the equivalent money to be spent on more general scientific education. In March 1964 a good deal of squabbling about this within the government ended in the arrest for corrupt practices in the administration of the CNEN of Felice Ippolito, who was regarded as the protégé of Riccardo Lombardi, then editor of *Avanti!* Ippolito's became a *cause célèbre* for he claimed to have been merely trying to cut through the bureaucratic labyrinth. In June, however, he was condemned to ten years' imprisonment and Lombardi was soon obliged to abandon his editorship.

In the summer of 1968, that is to say towards the end of the Moro period and late in the first Five-Year Plan, the Italian Government got to work on its further plans for nuclear development. Plants were to be constructed so that Italy should have 8,450 megawatts of nuclear power by 1980. Already in January 1968 ENEL had been offered a fourth nuclear power station. This was to be built by a new IRI company called the *Società Ansaldo Meccanica Nucleare* at Genoa which was also to construct the much talked-of Cirene prototype reactor for the CNEN. At about the same time ENI (or rather AGIP) was entrusted with a new activity, to prospect for uranium and work on it. As one would expect, ENI was instructed to co-ordinate its new nuclear activities with those of IRI in order to create one Italian nuclear organisation essentially based on Genoa. This was another step in the Socialist direction. State holdings were building up, though still not half-way. In so far as ENI and IRI acquired control over Montedison, this meant that over 36 per cent. of Italian enterprises with an annual turnover of at least ten milliard were under state control, but all these relationships seem deliberately to have been kept vague. The employees of the state concerns were engaged on particularly good terms so that even in 1969 and

1970 they were more hesitant to strike than other workers; they belonged to a special trade union called *Intersind*.

With its direct governmental authority ENEL unfolded very considerable activity in southern Italy, in which, incidentally, Cefis, like Mattei, showed no particular interest. Nothing, after all, could be more important in the interest of southern development than the supply of electricity. Based on Naples, by the end of 1969, with every kind of generator, ENEL's annual productive capacity in the south reached ten milliard kWh, and this is to be nearly tripled by 1976. During 1969 over 8·2 milliard kWh of electric energy was distributed in the south, 60 per cent for industry, commerce and agriculture and 40 per cent to private people – ENEL itself paid 20 per cent of the cost of rural electrification. From 1963 to 1969 ENEL had created installations in the south costing 29 milliard lire; they were distributed across the area to include Brindisi, the Gargano in Apulia, and many places like Sibari, hitherto without economic hope or history, at least in modern times. It would be tedious to make a list of all the hydroelectric and thermoelectric plants constructed for instance in Sardinia. The 29 milliard were provided by the *Cassa per il Mezzogiorno* (thus excluding Sicily) and the second agricultural plan or *Piano Verde*. Here again the number of jobs created was relatively small.

Another of the most profitable of Italy's economic activities continued, in spite of Spanish and Yugoslav competition, to be tourism. This also was controlled by the Italian state and included in its planning. For 1969 thirty million foreign arrivals were allowed for, whereas 31,232,000 foreigners actually appeared, the biggest national group from West Germany. There was also a considerable increase in Italian visitors to hotels and motels in Italy. This industry, which brings into the country, in a good year for both, more than the great motor-car industry with which its activities naturally interrelate, suffers from Italy's characteristic difficulties. For one thing it had been hoped all along that tourism would be regionally planned, but until 1970 the regions only existed on paper and even in 1970 they were chiefly engaged in organising themselves. A second problem is regional in another sense, that of the competition between the northern regions, the central ones and those in the south. Much bitterness over this, as over most questions, is felt in the south because it has fewer roads, railways and airports; plans have now been made for new airports at Naples, Bari and Agrigento. The basic trouble remains that southern Italy is much farther from the homes of most foreign visitors, a geographical fact which cannot be altered. Until the fifties there was an extraordinary shortage of hotels south of Rome with the one exception of the area of Naples in Campania. The completion of the *Autostrada del Sole* from Milan to Reggio Calabria – there are plans for a bridge or tunnel to Messina – makes the motorists' journey southwards pleasant but extremely expensive, since there are road-tolls to pay and the cheap petrol coupons for foreign visitors easily run out before they reach the real south. And then Milan, Venice, even Florence, are on the way to somewhere, but Calabria is not.

Visitors came to Italy partly to eat her fruit and drink her wine. Agriculture did not export much – indeed there was never enough wheat in Italy to make all the spaghetti – but Italy's export of fruit and tomatoes was important to her. In 1969 she exported more grapes than any other European country and for some time she had exported more wine than any country in the world. Membership of the European Common Market was particularly valuable in this connection, whereas in the steel industry and others of the kind it increased competition from the other five Community members.

The characteristic things about Italy's economic development in the second half of the sixties were not only the greater activity of the state, mainly through concentrations, but the growing power of the directors of the state concerns. In the fifties the old-style industrialist held his ground; in the sixties the next generation of Agnelli or Pirelli made concessions to their workers while beginning to negotiate mergers like Pirelli's with Dunlop abroad or with IRI companies at home. But this generation was to a noticeable extent overshadowed by the men like Cefis who were said to be more powerful than the ministers for whom they worked. Another factor in economic development, which was not peculiar to Italy but mattered more in her circumstances, was that the key and state industries tended to employ relatively few people, fewer as time went on. Neither ENI or ENEL provided very much employment, and the better they functioned the less jobs there were likely to be, while tourism offered erratic seasonal work. It was pessimistically reckoned by the economic

newspaper *Il Globo* that between 1963 and 1968 there was a
3·8 per cent reduction of all employment including agriculture –
1963 had been a bad year and 1968 a peak one. This was one
big factor in the social explosion that began to take place in
the overcrowded universities two or three years before the industrial strike fever of 1969–70, in a period of maximum prosperity.

Between 1965 and 1969 still the most conspicuously successful
Italian industry was the production of motor-cars, in spite of the
fact that Italy produced fewer cars then than West Germany,
Britain and France. The popularity of motoring and the fame
of the constantly changing Fiat models made it so. Fiat[1]
remained by far the biggest car producer, constructing in the year
1968 roughly two million of every kind of motor-car, tractor
and lorry; well over half were exported. The average annual
increase in production was 15 per cent. A great change in Italian
life was the substitution of tractors for the traditional farmer's
oxen in the centre and south of the country. Nearly 600,000 tractors, mostly made by Fiat, were in use in Italy in 1968. So powerful
did the Fiat people feel in 1968 that they began to negotiate to buy
up the whole Citroën concern, but the French Government intervened so that Fiat was only allowed to acquire 15 per cent of the
Citroën shares, although by 1971 Fiat looked like taking over the
whole of Citroën after all. By this time Fiat had launched its
venture of car construction in Togliattigrad on the Volga – there
was no little irony in this association of Fiat with Togliatti.

Although much smaller than Fiat, Alfa-Romeo, the motor
concern controlled by IRI, flourished too, and there were in
addition Lancia, Innocenti, Ferrari, Autobianchi and Maserati.
But within a year Lancia had joined Fiat, and Maserati and
Ferrari had been half taken over by Fiat. A freak of fashion
combined with the labour troubles of 1969 – labour relations
had hitherto been good at Fiat – to incline Italians to buy foreign
cars, and in this case too the EEC facilitated the tendency.
Indeed in 1970 about 27 per cent of the motor-cars bought in
Italy had been made elsewhere. The fortunes of the rubber firm
Pirelli depended to some extent upon those of Fiat, and it was
not surprising that negotiations for a merger with Dunlop were
pushed forward when Fiat's difficulties began in 1969.

[1] Although Fiat remained nominally independent, it is obvious that it
was entangled with much government activity.

In general it can be said that in the later sixties Italian refrigerators, television sets and sewing-machines continued to stand up well even to West German competition and also to the growing competition – not so direct – of Japan. On the other hand the fortunes of the Italian textile industry were very uneven and often unhappy. Building and shipbuilding were erratic, labour in the ports obstructing official plans for the different ports to specialise; Genoa, Venice, Naples, in that order, were the most important, though, remarkably, Augusta in Sicily was by 1967 importing more goods – i.e. crude oil – than Venice or Naples. The bulk of all this economic activity was still essentially based on northern Italy. The tax incentives of 1970, however, brought Costa to transfer his merchant fleet to Naples.

The economic history of central Italy in the second half of the sixties again proceeded at a moderate pace and satisfactorily. The Emilians and Tuscans did not have to go far from home to new industries, or to bring strangers in for them. They developed their own natural resources and activities into small-scale factories for the canning of fruit and vegetables. Machinery developed from agricultural implements to household goods and electrical equipment. In Bologna the cobblers' sons started shoe factories, the candlestick-makers' families set up furniture factories; clothes were made in Reggio nell' Emilia. There was only one big chemical factory, run by ENI near Ravenna; at Ferrara there was a smaller one. By the end of the period not quite 45 per cent of the Emilian population worked in industry, the figure in 1951 having been 25 per cent. The former *mezzadri*, who had often made wine, mostly combined in a co-operative which became a wholesale exporting affair. Italian statistics always give a high figure for people working neither in agriculture nor industry, but commerce and other things, and the statistical denomination of many central Italians must have been fairly erratic. Indeed Italian statistics, if not others, vary considerably according to the use of different criteria.

In central Italy, too, tourism brought in big sums, partly on account of the beauties of Florence and Siena, partly because of the 'development' of the Adriatic coast into a chain of resorts in the Marche. The southern portion of the Marche and Umbria were less happy; Umbria in particular suffered from the desertion of its difficult highland farms and there was a strange lack of 'development' of, say, Lake Trasimene, and a failure to develop the small-scale industry that flourished so thoroughly just to the north. Cortona was an indicator of what would be found south of it, though not in Perugia. Did the co-operatives of central Italy, especially of Emilia, work well because many of their members were Communists, or did they easily call themselves Communists because they were used to co-operatives? Logically the Communist mayors should have pressed for large-scale industrialisation, but they were as likely as not to wish not to spoil their countryside; even after John XXIII many central Italians voted Communist chiefly because they still hated the priests. Here in central Italy, unlike the North, it was possible to forget the south and to fail to understand that Italy could not come to terms with itself until the south had been truly developed.

It has been seen that the problem of the south of Italy was pronounced by Colombo soon after he became Prime Minister in August 1970 to be the very problem of Italy herself. His speech referred to above in its political context was made at Bari on 10 September in the presence of the economic potentates, Di Cagno, Sette and Agnelli, but not Cefis. De Gasperi's attempt in 1950 to solve the problem had not been without results. Great industrial plants belonging to the biggest concerns and to IRI have been established at Bari, Taranto, Brindisi and in many other places as well as in Sicily and Sardinia. At first southern labour was too unskilled to be taken on, but real efforts have been made to train and employ southerners. By 1970, however, the result was that there were industrial oases (if that is the word) at key points and especially at ports – 'cathedrals in the desert' they have been called – but that the contrast between the oases and the squalor of the still undeveloped parts was more striking than the old difference between north and south. Moreover, the high birth rate and number of underemployed in the south has so far proved unconquerable; in spite of the new industrial jobs in the south the exodus to the north of Italy has continued. Agriculture in the south which is typified by the cultivation of olives has been rationalised too, so as to offer less employment. There have been several unfortunate consequences. After national production picked up in again in 1965 the rate of its increase continued to be higher in the north than in the south so that, irregularly, it is true, the gap between the two tended to widen.

The exodus from south to north had been re-stimulated after the 'miracle' by the fact that it did require proportionately bigger investment to start up new concerns in the south than to expand northern ones. Further investment in the north was also encouraged by the fact that it came to be felt in the south that men between twenty and fifty had no initiative unless they looked for work in the north. Altogether the process of industrialisation encouraged developments which left southern villages with no men but the old ones – a bleakly demoralising condition.

The hitherto unsolved problem of southern Italy was illustrated by an article in the *Neue Zürcher Zeitung* by its economic correspondent in Italy on 16 September 1970. He was writing about the opening of the Trade Fair at Bari, which he too had visited. This fair had been going on for half a century but it was only in the quite recent past that it seemed squarely based on the capital city of Apulia. Bari, together with Brindisi and Taranto, is now referred to as one of the towns of the 'southern industrial triangle'. Fiat alone had already invested 86 milliard lire in Apulia, that is in Bari, Brindisi and Lecce, and was about to invest a great deal more, directly offering 19,000 jobs and indirectly about double that. A notable sum had been invested by Alfa-Romeo (belonging to IRI) in a motor factory near Naples, *Alfa Sud* at Pomigliano.

The correspondent of the *Neue Zürcher Zeitung* pointed out that between 1951 and 1968 income per head had doubled even in the south, and thanks to the Five-Year Plan of 1966–70 investment in southern industry had increased by 18·8 per cent whereas the increase in the north and centre had been only by 8 per cent. Production in the south had increased absolutely, too, but significantly the annual increase of jobs in the south was only 107,809 whereas in the centre and north it was 268,519. For the *rate* of production had lagged behind and the southern share of the national income had declined (not steadily) between 1951 and 1968 from 24·5 per cent to 22·5 per cent, especially so between 1967 and 1968, for in 1967 it was 23·4 per cent. Roughly speaking, wages in the south were half what was paid in the north.

A newer governmental Ente, that for the *Finanziamento dell'Industria Manifatturiera*, called the EFIM, put great emphasis on its southern activities. Although it had been legally constituted in 1962 and its creator and then president, Pietro Sette, had

been a friend of Mattei and belonged to that category of energetic
director, not much was heard of EFIM for some years. In 1968,
however, it invested 19 milliard in the south, in 1969 25 milliard,
but in 1970, 101; it was intended that by the end of 1975 this
figure was to soar to 608 milliard. Its policy was to create smaller
industrial units which could offer high employment. It built,
inter alia, glass and tyre factories, for instance in the Abruzzi, to
provide for the needs of Fiat and Alfa-Romeo in the south without
recourse to the north. EFIM also contributed to the building of
hotels and other touristic development. It started cement works
in remote Lucania and Calabria and the development of
aluminium in Sardinia; on the list of southern places where it
established factories, names like Matera, once a symbol for
despair, were to be found. IRI had already established its Finsider steel works at Taranto in 1965. At the Bari Fair in 1970
the southern Dalmine steel plant, held also by IRI, displayed its
wares under the motto 'Acciaio come produzione, acciaio come
consumo' – steel was to be *used* here as well as made. Again it
was being attempted to make the whole complex of southern
industry more independent of the north. In a sense planning was
having a divisive effect without increasing production above the
rate of the miraculous years.

The whole business of planning has hitherto been defeated
before everything by the monstrous emigration from south to
north, some six million people, it is estimated, since the end of
the war, workers whose families sometimes followed them. That
a couple of million people from the south should have gone north
before the programme era, i.e. in the miraculous period, does not
seem so incredible, but that even more should have done so after
1964 seems extraordinary. The fact that Switzerland introduced
limitations to the number of foreign workers it accepted is a quite
inadequate explanation. The flow had not subsided at all even
by October 1970 when further jobs advertised in the north were
being pursued. At about that time the *Centro Orientamenti
Immigrati* (COI) held a meeting in Milan;[1] this was attended
by many of the new regional authorities who hoped to tackle the
problem. The figures quoted showed that Milan, Turin and
Genoa had continued their fantastic and alarming growth. Milan
was stated to have roughly 1,700,000 inhabitants by this time,

[1] *La Stampa*, 11 Oct 1970.

although only half a million of them were natives of the city. In the year of the census of 1951 Turin had counted 719,000, but in 1970 its population was up to 1,152,000. In Genoa things were not so excessive, the increase being only from 688,000 to 843,000. As long as this flood is not checked, the north of Italy can never arrive at any kind of equilibrium, even a flexibly modern one. None of the social problems of the day – education, the health service, housing, town-planning – can be dealt with to any purpose. The flow to the north does not contribute to Italy's solvency as does the emigration of Italian workers to foreign countries. Finally it has meant that investment in the south has not been fully exploited. As usual great things were hoped from the regions, and it was proposed that the CIPE (*Comitato Inter-ministeriale per la Programmazione Economica*) should be trans-formed into an inter-regional committee.[1] This decision was taken a year after the peak of industrial production had been reached and the records achieved then destroyed by wholesale strike action. No one really knew how far the plans had been fulfilled, but they had all been distorted by the exodus to the north. Although, unlike Rome, Milan had had a splendidly successful underground railway since 1964,[2] its commuter services, which directly affected the immigrant workers, were inadequate.

For already in the 'hot autumn' of 1969, the fine growth, in spite of all obstacles and inadequacies, of Italian industry was abruptly halted by the outbreak of strikes, official and other, in all directions; Fiat, where labour relations had been relatively good, was particularly badly hit. It has been seen that by this time a new generation of industrialists was emerging. During April 1970, shortly after Rumor's statement of Cabinet policy on 7 April, the *Confindustria* chose a new chief, Renato Lombardi, in the place of Angelo Costa, the shipowner who had been its president for thirteen years. Some ten days later Cesare Merzagora, who was far from young (having been President of the Senate), succeeded Valerio as the President of Montedison. Really younger men like Leopoldo Pirelli and Giovanni Agnelli, the latter now personally in command of Fiat, had shed the paternalism of their fathers; it

---

[1] Originally this was the *Comitato Nazionale per la Programmazione Economica*; see above, p. 45.

[2] The cost of 44 milliard lire was borne by the city of Milan; it had taken seven years to build.

was not nearly so difficult to negotiate with them. In due course a number of new contracts were made. Little was heard now of the 'participation' once so dear to Gramsci, but higher wages linked with greater productivity and shorter hours were gradually introduced, beginning with the contract made in these terms by their employers with the *metalmeccanici*, the well-organised metalworkers,[1] in December 1969.

Notwithstanding such agreements the leftist crusade to get all the most cherished reforms put into practice as the result of strike action gathered force. This was using strikes for a political purpose with a vengeance, and many workers were opposed to it. It was a matter of debate as to how many workers belonged to the unions – they themselves did not claim more than 40 per cent, even in Lombardy, for the three big organisations of the GCLI, CISL and UIL, and the figure generally accepted was five out of twenty million workers in the whole country. Moreover, to block production was to reduce the resources of the state, directly or indirectly, when in fact huge sums would be needed to transform the old 'mutual' insurance societies into a health service, to provide cheap housing and new school and university buildings, let alone grants for students. Rumor was made wretched by this contemplation and his dejection over it was surely one reason why he resigned in July 1970 after the unions – though not the UIL – had announced their general strike of twenty-four hours for 7 July. That strike was cancelled, perhaps partly because a large-scale settlement was in the offing between Agnelli and his people; they had been discussing it for three months and now on 8 July agreement was reached in Turin. Wages linked to productivity were to go up substantially and a forty-hour week was to be introduced by stages. It has been seen that an agreement of this kind had already been made with the metalworkers, and there had been other similar ones in other industries. Now, before the end of July itself, nearly as much was produced by Fiat as in the record month of July 1969. Fiat was the biggest private concern in the country and reflected the general position for the moment.

By the time Colombo was installed as Prime Minister early in August 1970, wage-earners had had time to discover that higher wages did not help them to find flats or doctors they could afford.

[1] See above, p. 79.

And so the talk of a general strike to force through the major reforms – health, housing, education – blew up again. The objective seemed commendable, although a general strike might well destroy the democratic state along with the economy. Again observers wondered whether the Communists were at last coming to the point of taking over. Colombo had the courage to look them in the face and demand from the people of Italy financial sacrifice as well as patience. He drew up the *decretone* of 27 August. This was conceived as a fiscal measure which would cancel the worst debts of the 'mutual' health societies while taxing all luxuries heavily. The Minister of the Budget, Giolitti (PSI), spoke of it as intended to increase social investment while hitting the really poor as little as possible. On 20 September 1970 these things were all in doubt, but by the end of the year the Italian Republic was still on its feet, laicised by Parliament's acceptance of divorce, and economically viable thanks to the acceptance of a revised version of the *decretone*. The latter made petrol still more expensive and indeed introduced something of the austerity of which the Prime Minister had spoken at the Fair at Bari. To the foreign observer Italy seemed to be growing so fast economically that the seams of her clothes were bursting; she seemed dynamic rather than austere. The final version of the *decretone* went a long way towards the implementation of the most vital reforms. In the early hours of 1 December 1970 the Chamber accepted this mini-budget, now linked with the divorce proposals, by 359 votes to 246.

During 1970 Italy's imports rose by 20 per cent and her exports only by 12½ per cent. The deficit caused by the *malattia inglese*, or English illness as the Italians called striking, was prevented by Carli's skill from creating disaster. For he again successfully brought exported capital home by promoting Eurocurrency borrowing by ENEL, the state railways, the Public Works Institute (*Creditopere*) and other bodies. The problem of the south was not yet near solution. At the end of 1970 it was reckoned that 26 million, that is nearly half the whole population of Italy, were inadequately housed. But 85 per cent of the people living in the north were properly housed and only 17 per cent of those living in the south. These figures were given in *The Economist* of 20 March 1971 without reference to the good conditions in central Italy.

The Italian Government, through the *Cassa per il Mezzo-giorno*, already paid up to 40 per cent of the employers' share of social insurance in the south, and other incentives were sought for all the time. There is reason to hope for the real 'take-off' in the south quite soon.

10 Southern Close-ups and the Young

Sicily has a fantastically elaborate and varied history; wave after wave of civilisation has passed over the island. Yet until the present its history has more often than not been the history of decline. This is perhaps the reason for the evil traditions of Sicily; the Sicilians might indeed be termed an anti-social society, and in so far as their habits have spilt over into the mainland this has been deplorable. For Sicily is like an evil parody of Italian society, or rather it was so until the economic miracle shook Sicily to its pagan foundations too.

In the Fascist period there had been some sharp police action against Sicily's great curse of the Mafia whose leaders went into exile in America. Nothing serious, however, was done to banish the terrible poverty of the island, for Mussolini preferred to decorate Rhodes. When the Anglo-American forces invaded and liberated Sicily in 1943, the Americans with their DDT put an end to malaria, but they allowed the *mafiosi*, who had taken refuge in the United States, to return to Sicily. This was a gloomy prelude to the introduction of the new Italian Republic which countered a flickering separatism there by granting Sicily regional autonomy. The Mafia needs explanation. It was an inherited system of ruthless blackmail. For centuries clan-like groups had threatened others, who were unorganised, with any kind of punishment, including murder, if they resisted the wishes of the menacing groups. Once the Sicilian Italo-Americans were back, Sicilian life could not be imagined without this, certainly until 1962. There was nothing like it on the mainland. It must be added that the Mafia was powerfully established only in western Sicily; in the eastern half of the island its influence was negligible; it has been said that this is due to the better water-supply in the east which has militated against economic backwardness.[1]

[1] P. Sylos-Labini, *Problemi dello sviluppo economico* (1970), p. 179.

The Mafia was like quicksilver to handle – or to describe. As Denis Mack Smith has recently written:

> The Mafia in some villages might be just four or five groups contending for civic office and economic power: their number, composition and political colour might even be continually changing, and their involvement with crime could be a secondary matter or non-existent.
>
> Many towns and villages, however, were the scene of gang war accompanied by political assassination and drug smuggling on a scale never before known.[1]

The capital city, Palermo, in the western part of Sicily, was highly susceptible to the influence of the Mafia, and the political life of the autonomous island was distorted by a *mafioso* like Genco Russo standing as a Christian Democrat deputy in Palermo and holding public office. Palermo was inevitably the centre for Sicilian government contracts and official permits, and for years, although the existence of the Mafia was often denied, it grimly if erratically affected the government of the island, impeding positive achievements.

For the *mafiosi* seem to have descried their interest in the retention of the *status quo*, huge neglected estates, great poverty and evil conditions for the majority. These made people more susceptible to bribes and threats, whereas, if people worked for decent wages in industry or on the land and organised themselves in unions, they were likely to become more independent. De Gasperi's land reform included Sardinia but not Sicily. However, the Mafia notwithstanding, Sicily introduced its own agrarian reform, and neglected *latifondo* land was expropriated and redistributed to small owners. Indeed model villages were built recklessly in Sicily and then perhaps not used. The central Italian inclination towards the formation of co-operatives was not to be found here: even in eastern Sicily 'excessive individualism', as the saying went, prevailed. Gradually, nevertheless, some progress was made in irrigation and in the transfer to wine and tomato-growing, instead of wheat which provided little work and exhausted the soil. Sicilian oranges and lemons at first found it difficult to compete on world markets, and the sulphur mines, which had once virtually monopolised sulphur production in the

[1] Denis Mack Smith, *Modern Sicily after 1713* (1968), p. 539.

whole world, were now quite unable to stand up to modern times.

In 1953, however, there occurred the beginning of an economic revolution in Sicily: what gas did for Italy – the north at any rate – oil did for Sicily. It did more than that, for it made Sicily into Italy's chief oil-well. The Sicilian Government, which was independent of Mattei, had offered good terms to foreign oil companies. It was in 1953 that the Gulf Oil Corporation struck oil near Ragusa in the south-east and production began there at the end of 1954. Not long afterwards ENI struck oil on the south coast at Gela. Four refineries were soon built to deal with Sicily's oil on the island, and large quantities of crude oil were imported to them from Russia and the eastern Mediterranean. Such developments contributed to the fact that between 1953 and 1963, for example, the tonnage handled in Sicilian ports multiplied by about six, and imports to Augusta even began to outdo Venice and Naples, though not Genoa.[1] Then, early in the sixties, gas was discovered at Gagliano in the centre of Sicily, and later at Bronte, and the necessary pipelines were built. The development of potash deposits added to this story. Thus Sicily, once the granary of the central Mediterranean, became not its particular oil-well but one of its oil-refining centres with plenty of subsidiary economic activity: instead of being a drag on Italy the Sicilians could boast with justification that they were now contributing to Italy's prosperity. The oil strikes on the Italian mainland had proved a disappointment to Mattei; he could but welcome the Sicilian contribution, although it was not as firmly under his control as if Sicilian autonomy had not existed. But all the same Sicily had become an asset even to Mattei. It should be added that Sicilian oil was not of high quality or suitable for all purposes; moreover in the later sixties production declined somewhat.

The effect of the establishment of the oil industry in Sicily was a mixed one. Standards of living rose appreciably but again the oil industry was not one that provided much employment or sense of social co-operation. The sources of oil were in the eastern half of the island where the Mafia had little say. Its influence continued to affect the western half of Sicily and therefore politics in Palermo. Originally the Mafia had favoured Sicilian separatism, but it switched, as soon as that cause was lost, to an anti-Left position in Italian politics. Just after the war the brigand

[1] See above, p. 97.

Salvatore Giuliano who was vaguely associated with the Mafia opened fire on a defiant leftist May Day meeting in 1947. Giuliano was killed in 1950, but the Mafia's feud against the Left continued. The Communists in particular made efforts to expose the *mafiosi* – it was characteristic that the office in Palermo of their paper, *L'Ora*, was blown up as late as 1958.

Thanks to the Mafia the corruption which was so much talked about on the mainland, but was probably only widespread in Sicily, continued to prevail. Sicilian politics were labyrinthine and typified by the performance of a renegade Christian Democrat leader called Silvio Milazzo; in October 1958 he became Chief Minister with – against all the rules – support from Monarchists, *Missini* and Communists. This brought a reproof both from Ruffini, the northern Archbishop of Palermo, and from the Government in Rome, but Milazzo was defiant.

After this, in spite of the exodus of workers to the north, even Sicily showed signs of greater stability. In the spring of 1961, under the leadership of a Christian Democrat called Giuseppe D'Angelo, it established a pioneer *Centro-Sinistra* Government. The Sicilian authorities had denied the existence of the Mafia in order to avoid taking action against it. Now in 1962, not long after D'Angelo had taken office, the Governments, both in Rome and Palermo, agreed to set up a commission with wide powers to investigate the Mafia and consider remedies. In 1963 when Rumor was at the Ministry of the Interior under Leone, he ordered firm police action against *mafiosi* who were begging off by producing membership cards of the Christian Democrat Party. People did become less afraid to give evidence against them. When in August 1966 a number of new houses in Agrigento collapsed,[1] it was said quite openly that the building contracts had been in the hands of *mafiosi*. Deputies and senators in Rome were no longer afraid to name the Mafia in parliamentary debates, but the commission under Senator Pafundi was unable to suppress its activities which flowered again a little later through trafficking in drugs. Indeed the Mafia during the sixties seemed to have become much more of a commercial organisation; it had, after all, moved with the times. Nevertheless in the night of 19–20 June 1970 a herd of some 170 sheep was killed near Caltanissetta, just in the western half of the island, in the most traditional Mafia

[1] See above, p. 63.

fashion. In the later sixties Sicilians, incidentally, again made up a substantial proportion of the southern workers who moved to the north.

Speaking in general terms there can be no doubt that by 1970 life in Sicily had become more normal by Western European standards. To this extent the hopes of Salvemini, whose wife and children were killed in the earthquake of Messina in 1908, had been fulfilled, and the efforts of Danilo Dolci – whom the Mafia never bothered to touch – to reform the island seemed more justified. Against this there had been the minor earthquakes at Gibellina and Trapani in 1968 which did not attract European sympathy, and which were feebly handled by the Sicilian Government whose financial difficulties were desperate. The unending crisis in the Middle East was like a threatening black cloud, and the revolution in Libya in September 1969 led to the expulsion of the Italians and made the position of Sicily more frightening. A pro-Russian Maghreb plus Libya made Communist pressure, whether internal or external, greater.

SARDINIA

Sardinia, too, had an intricate past, though not quite comparable with that of Sicily. It had been ruled by the Romans and Byzantium and its dialect is particularly close to Latin. For four centuries the Spaniards neglected it, and the efforts then made by the House of Savoy after 1720 to catch up with history there could not get very far. After Italian unification, or rather at the beginning of the twentieth century, it seemed possible that Sardinia would make a substantial contribution to Italy's industrialisation since it contained considerable deposits of zinc and lead and a lot of lignite or poor-quality coal. When Istria was lost to Yugoslavia at the end of the Second World War this was virtually all the coal Italy contained. But then came Mattei and the gas and the oil, and Sardinia became the most burdensome of all the southern regions with the possible exception of Calabria. Unlike Sicily Sardinia was at least not overpopulated, having in 1970 only about one and a half million inhabitants to Sicily's five million, occupying a fairly similar area.

In 1946 the American Rockefeller Foundation abolished malaria in Sardinia with DDT. Three years later the island was given regional autonomy. Then, until the sixties, nothing more happened except banditry, in itself a sign of nothingness; there was nothing so subtle as the Mafia in Sardinia – the bandits in Sardinia were outlaws whereas the *Mafiosi* in Sicily often held civic positions. Sardinian politics did not diverge far from the national pattern, the Christian Democrats always being the biggest party. The Sardinian Left was strengthened a little by the survival only here of the Party of Action. This survival was due to the heroic old anti-Fascist *Azionista*, Emilio Lussu, born in the *paese* of Cagliari in 1890, who was made an Italian senator for life by the Republic. Lussu had been a friend and colleague of Parri; he returned from exile in France in 1943, working clandestinely till the Germans had gone. Now the second-in-command of the PCI in Rome is the Sardinian Enrico Berlinguer.

It was not until 1960 that a plan for the economic take-off of the region of Sardinia was launched and at first it proceeded very slowly. An attempt was made to use the antiquated lignite mines at Sulcis for a power plant, but this failed and the Sulcis mines, like the Sicilian sulphur mines, had to be closed down. It has been seen (p. 91) that during the sixties the *Società Italia Resine* or SIR of Milan had built a petro-chemical plant at Porto Torres in the north-west of the island offering about 8,000 jobs. The guiding spirit here was a man called Nino Rovelli and there was co-operation with the smaller company of *Rumianca*. Synthetic fibres and paper were produced. Petrochemical plants were established also at Cagliari and in the centre of the island. After 1967 things really went ahead: reafforestation was seriously undertaken with the transplantation of suitable trees from New Zealand. Orgosolo, which had been the bandits' centre, began to go out of that business when in March 1968 Graziano Mesina, the leading bandit, was caught. Meanwhile ENI was building a huge petro-chemical plant at Ottana. Tourism leapt ahead, especially along the Costa Smeralda which was developed by the Aga Khan.

Sardinia has two universities, one at Sassari, the other at Cagliari: strangely one never heard of trouble from their students. And in spite of development and a small population, Sardinian young men added to the flow of southern labourers to northern Italy.

CALABRIA

Ever since 1870 large numbers of the former subjects of the Kings of Naples had felt an exasperated bitterness against what they regarded as their annexation by the Piedmontese. It was particularly in the deep south, Lucania and Calabria, that neglect by the new kingdom of Italy had justified such feelings. Fascism had cared little about them, but after that, when the Republic was founded, the voices of old *meridionalisti* like Salvemini and Zanotti Bianco (himself Piedmontese) seemed much more likely to be listened to, and a member of the next generation, Professor Manlio Rossi Doria, was among those who urged De Gasperi to help the south. It has been seen how before 1970, in spite of many genuine efforts, the south never quite managed to 'take off', to get off the ground economically. Nevertheless Bari–Brindisi–Taranto had become a recognisable industrial triangle and Lucania had benefited a little from their relative proximity. Sicily had its own erratic Government with erratic success, and quite a lot of capital was gradually directed towards Sardinia. Calabria, however, always came off worst and believed itself forgotten. None of the outstanding political leaders seemed to speak on Calabria's behalf: Moro was from Bari, Colombo from Lucania – it was true that Fanfani's mother had been Calabrian. If one goes through the various projects for southern development, Calabrian bases appear relatively seldom and there is no Calabrian port to rival Taranto and Brindisi, let alone Augusta in Sicily, no Calabrian industrial triangle to rival that of Apulia.

In 1970, with the prosperity of others in sight, the south Italians were no more reconciled than they had ever been. If the regions were to do the planning now, one could not help but fear that their rivalry might resemble that between the city-states before the great foreign invasions. At first the jealousy was within the regions, competition between cities wishing to become the regional capital. In fact in southern Italy poverty, anxiety and ignorance brought about frantic competition between the towns in order to be chosen as regional capitals. In the Abruzzi there was jealousy between Pescara and L'Aquila and riots at Pescara when towards the end of June L'Aquila was chosen – indeed the prefect of L'Aquila was chosen to be the central Government's representative in the region of the Abruzzi. (Fortunately Perugia had no rival in Umbria.) It was naturally in Calabria that this kind of situation became most alarming. Reggio di Calabria is its best-known town, closely related to Messina geographically and in other ways. In order to help Cosenza, which seemed less well placed, a new university was founded there; this could perhaps be justified in that Messina had long had its own university from which Reggio could benefit. Then in July 1970 Cantanzaro, which already had the Court of Appeal, was proclaimed the regional capital. This led to seven months, beginning on 14 July 1970, of what became civil war at Reggio di Calabria. Perhaps the most bitter grievance of the citizens of Reggio di Calabria was the feeling that, whereas Catanzaro had a powerful political boss in the PSI leader, Mancini, and Cosenza in the young Christian Democrat Minister of Education, Misasi, they had no one to get jobs and contracts for them.

The outbreak began just after Rumor's resignation. Northern journalists were horrified by this senseless revolt of mainly very young people who suddenly turned into sophisticated rebels. Almost all the male population between twenty and fifty had gone to northern Italy; their juniors seemed to live up to and compete with the *contestazione* or challenge of northern students to authority, for southern students had not demonstrated. Naturally the local *Missini* and the local *mafiosi* – for, strangely, *western* Sicily was believed to spill over into Reggio – stirred up the trouble for all they were worth. It was utterly destructive and pitiful, one witness describing it as a senseless *jacquerie*. But Calabria, it has been seen, was at the end of the road, more so than Sardinia or even Lucania; it had no major port, no airport, still in 1970 almost nothing, and nothingness breeds anarchy. No one knows yet how this can end. It would be quite misleading to suggest that Reggio di Calabria represents the whole south where the achievements in the last quarter of a century have been about as great as the failures. The *Cassa per il Mezzogiorno* was renewed in 1965 for fifteen years; it has done great things, but after its early years it was too readily taken for granted, and tended to come to the rescue of bankrupt local authorities rather than save itself for special operations.

THE YOUNG

Young Italy is certainly in ferment, political ferment mainly. The young there have less time and opportunity for drugs and sex than their Anglo-Saxon equivalents. Conditions in the universities are far worse than they are in Britain. Everyone who has passed his final school examination has a right to study, so that the overcrowding is indescribable and there are very few grants or scholarships. The professors, as in France, have too many outside interests; it is characteristic that Fanfani and Moro are university professors who periodically retire to their universities. Some kind of Marxism is almost taken for granted among the more intelligent and active students; perhaps Marx satisfies the thirst of the Italians for political doctrine, for ideology.

It was mainly by chance that the students of Italy rebelled *en masse* against authority at about the same time as those of Czechoslovakia and a year before the French students rose up. In Czechoslovakia they were part of a national and liberal movement against Communism of truly almost nineteenth-century type, a movement later crushed by the Russians. In Italy there was every motive other than the Czechoslovak one. The Italian students really had very serious practical grievances. It has been seen that the universities were appallingly overcrowded, the teaching old-fashioned and impersonal. In fact there was virtually no devoted tutoring as in Britain, but this was practically impossible because the number of students was seven times what it had been in 1940 whereas the teaching staff was only three times as large. In 1968 the University of Rome alone had 63,000 students.

For idealistic students with the political or religious fervour of Italians, 1967 was indeed a year of despair. Pope Paul VI was obviously no true successor to John XXIII with his love towards 'all men of goodwill'. Moro and Rumor failed to interest them and they despised the Christian Democrat Party as narrow and greedy. Most Catholic students, it has been seen, were looking towards the Ingrao wing of the Communists. The more politically-minded leftists were disappointed by the divisions among the Socialists and what seemed to them the conservatism of the Communists, perhaps the authoritarianism of both Communists and Christian Democrats. As for the *Missini*, it seemed to most Italian students that it was an unbearable breach of the constitution that they were allowed to organise themselves at all. Of course a few students found it very exciting to join the *Missini*. In May 1966, indeed, there were violent clashes between *Missini* groups and Socialist students and one of the latter, Paolo Rossi, was killed. The *Missini* were in fact fiercely oppressive towards the high proportion of leftist students, leftists of all kinds. There were many who followed Peking and were called 'Chinese': the *Manifesto* group in the Chamber more or less represented them.

It was, however, in the autumn of 1967 that the students of Italy began to impose themselves upon the political scene by occupying the universities. The authority which most affected the students was, of course, that of the rectors and professors of their universities; the universities, like most schools, were directly under the state and to the students the most obvious illustration of 'something rotten in the state of Denmark'. The Chambers were intermittently discussing the reform of the universities when the student occupations began. Over 800 arts students occupied the buildings of the University of Turin in November 1967 as a protest against the authoritarianism of the university. Between then and February 1968 the rector, who, as a state official, had the right to do so and had no proctors, called in the police five times to expel the students, but they always returned in greater numbers.[1] Many of the students calling themselves the *Movimento Studentesco* were followers of Marcuse and wanted a *contestazione globale*, a challenge to the whole structure of society. In Turin and elsewhere they demanded a general assembly of students which should create study groups to examine what reforms were necessary. They demanded student grants and the extension of the teaching staff through arrangements for the more advanced students to teach the younger ones. A minority of the professors supported the students who also demanded a say in judging student performance. The University of Rome was occupied by the students in February 1968; indeed occupation was so general that the Director's quarters in the *Accademia delle belle Arti* in Venice were occupied by art students. Meanwhile the Chambers were dissolved in the spring and the legislation planned for the reform of the universities was lost with this dis-

---

[1] Stuart Woolf on 'Student Power in Italy', *New Society* (Apr 1968); also R. Boston on 'The Italian Chaos' (May 1969).

solution. The ferment continued intermittently. Sometimes the students were strangely fierce, flinging the now meaningless word of 'Fascista' in the face of professors who, a generation before, were partisans fighting against the Fascists and the Nazis.

At about Christmas 1968 the present writer listened to the political confession of faith of a schoolboy of fifteen. He lived at home, as nearly all continental schoolboys do, the son of leftist but undogmatic parents who had sent him to one of the rare élitist schools in Rome: he was great friends at school with a young prince belonging to a famous Neapolitan family. He had little knowledge of the working class unless one counted as such the retainers of the young prince's family – at his own home there was at most a Swiss 'au pair' girl. But he had just 'gone Chinese', as it was usually called in Rome, and declared that only the *classe operaia* was pure enough to be preserved – everything else must be destroyed, yes everything. Apparently most of his school friends and some of his teachers held similar opinions. Much the same was true in other cities. In fact the university students had by now become less active. When they clashed with the police Pier Paolo Pasolini was on the side of the police because he said that the students were 'bourgeois' and the police 'proletarian'. I suppose my young friend of fifteen agreed with him though I had no opportunity to ask him. Colombo's Government is to try to tackle all the educational problems together with the rest.

11 Literature and the Arts

There has been a great flowering of the arts since the end of Fascism and the war in Italy, particularly in literature, the cinema and the theatre. Alberto Moravia emerged quickly as one of the leading novelists with *La Romana* in 1947 and *Il Conformista* in 1951. Already in 1945 Carlo Levi, a painter by profession, published a remarkable description of his banishment to Lucania under Mussolini. The book was called *Cristo si é fermato a Eboli*, and it helped to focus attention on the problem of the south. Later, in 1959, Tomasi di Lampedusa's *Il Gattopardo* focused attention on Sicily.

There has been a wealth of novels, the prevailing style of which is sometimes called neo-realist, sometimes naturalist, and is sometimes regarded as a late form of expressionism. One of the most distinguished and erudite novelists is an older man called Carlo Emilio Gadda, born in Milan in 1893. He has, like many Italian writers, written mostly short stories and essays; one of his most distinguished novels came out in 1963, *La cognizione del dolore*. Another outstanding writer is Vasco Pratolini, essentially Florentine and working-class, born in 1913. One of his best novels came out in 1947, *Cronache di poveri amanti*. The tragic figure of Cesare Pavese cannot be neglected, a collaborator with Giulio Einaudi in publishing, an eager translator from English and American and the author, just before his suicide in 1950, of a novel called *La luna e il falò*.

Italo Calvino is an attractive writer with magical powers of description, from Liguria, the youngest of all these and much influenced by his life as a partisan. He made his name with his first book, *Il Sentiero dei nidi di ragno* (1947). He was co-editor with Vittorini of the controversial literary magazine *Il Menabo*, now out of circulation. Elio Vittorini, born in Sicily in 1908, was in some ways most famous of all, author of *Uomini e no* (1945) about the Resistance, but he had published *Conversazione in Sicilia* already in 1941. He died in 1965.

Other first-rate writers of novels and short stories [1] are Carlo Cassola, Mario Soldati, Natalia Ginzburg and Giorgio Bassani. Cassola, much influenced by Joyce, as many of these people are, also fought with the partisans; now he often writes for the *Corriere della Sera*. His novel, *La ragazza di Bubè*, won the Strega prize in 1960. Natalia Ginzburg, who writes with a not dissimilar simplicity, is a quite fascinating writer. Her novel, *Le voci della sera*, appeared in 1961 and her autobiography, *Lessico famigliare*, describing the remarkable life of her family, the Levis of Turin (her sister was married to Adriano Olivetti for a time), in 1963. Originally she married Leone Ginzburg of Russian Jewish origin who perished in struggling against the Nazis; he, too, had been a close collaborator with Giulio Einaudi. *En secondes noces* Natalia Levi married Gabriele Baldini, learned in English literature, who died tragically, relatively young, in 1969. Elsa Morante, for a time married to Moravia, should be mentioned too as a distinguished novelist.

Mario Soldati is Piedmontese like Pavese and about the same age: his novel *Lettere da Capri* won the Strega prize in 1954. Giorgio Bassani was born in Bologna in 1916 and worked for some years for the publisher Feltrinelli for whom he discovered Lampedusa's *Il Gattopardo*. His own first book was published in 1945, *Storie dei poveri amanti*, but his *Il giardino dei Finzi Contini* in 1962 is perhaps the most famous of his works.

It would be wrong to ignore the Sicilian writer Leonardo Sciascia who is obsessed with the crimes of the Mafia as well as of the Catholic Church. Probably his best-known book is *Morte dell' inquisitore* published in 1964 and describing a terrible *auto-da-fé* in Spanish Palermo in the seventeenth century. Umberto Eco, born in 1932 and a reader at the University of Turin, is an avant-garde critic finding expression in the manifestos of *Gruppo 63* and in the magazine *Il Verri*.[2]

Lastly something should be said of Ignazio Silone who began life as Secondino Tranquilli, an orphan in the Abruzzi, born in 1900. In his youth he became a Communist and worked with Gramsci in Turin, but in 1930, when he settled in Zürich until

[1] See Raleigh Trevelyan (ed.), *Italian Short Stories* (1965).
[1] See below, p. 120.

1945, he broke with the Communist Party. The novels he wrote before the war were highly successful in German translation, but until a year or so ago the Italians did not recognise his merits. Recently, living in Rome, he has been co-editor with Nicola Chiaromonte of a review called *Tempo Presente* which is often equated with *Encounter*.

The outstanding publishing firms of Einaudi and Feltrinelli have already been mentioned. It would provide an incomplete picture not to speak also of Laterza of Bari, famous as Croce's publisher for many years, but still very busy since Croce's death. Mondadori of Milan blotted his copybook badly in Fascist days and has had to re-establish his reputation as a publisher. Alberto Mondadori, son of the founder Arnoldo, was born in 1914, and in 1950 successfully founded the magazine *Epoca*. In 1958 he won the Viareggio prize for poetry.

There has been an outburst of historical writing since 1945. The father of it, and indeed in a sense of the Republic, was Gaetano Salvemini who had lived in exile at Harvard during the major part of the Fascist period. He was back at the University of Florence before the end of 1945 and began to lecture by quoting the sentence with which he had left off some twenty years earlier. A remarkable history book came from a very different quarter, A.C. Jemolo's *Chiesa e Stato in Italia negli ultimi cento anni* published in 1948. Mario Toscano, in charge of the official Foreign Office documents now gradually being published, wrote many volumes on foreign policy; on his death he was succeeded by Giorgio Carocci who published in 1969 a part-history of Fascist foreign policy. The most gifted of all the historians was probably Federico Chabod, an Aostan who wrote *Le Premesse*, a history of Italian foreign policy from 1871 to 1896 based on the documents and at the same time a great work of art, published in 1951. He died in 1960 before he could write very much more for he was ill for some time. The most persistent historian of the next generation is Renzo De Felice who is in the course of writing a five- or six-volume biography of Mussolini: the first volume, *Mussolini: il Rivoluzionario*, appeared in 1965. A great deal has been written about Italy's intervention in the war in 1915, notably Brunello Vigezzi's *L'Italia dalla neutralità all' intervento nella prima guerra mondiale* which also appeared in 1965. The Communists have written a lot about Gramsci and the Resistance, including Roberto Battaglia on the latter subject. Paolo Spriano is writing a history of the Italian Communist Party of which the first volume appeared in 1967 and the second in 1969. Of course scholars have been writing about history before 1900 but less conspicuously.

The universities are making efforts here and there to catch up with the history of the twentieth century, but part of the students' indignation is caused by the slowness in doing so, as well as by the overcrowded lecture-rooms and the political preoccupations of some of their professors. Rosario Romeo, Nino Valeri, Franco Venturi and Leo Valiani, who in 1966 published an important book on the dissolution of the Habsburg Empire, are among other historians who have distinguished themselves. Among the Catholics, the Jesuit writer Giuseppe De Rosa has written histories of the *Popolari* and of Catholic Action.

The Italian press is extremely vigorous and varied, and many would say that the Italians just are journalists by nature. It is interesting that several of the most famous newspapers were founded before Italy was completely united, *La Stampa* of Turin for instance, in 1866, and *La Nazione* of Florence as far back as 1858 – the year of Plombières. The *Corriere della Sera* of Milan, founded in 1875, had a great period under Luigi Albertini as its editor until Fascism crushed it; after 1945 it re-emerged, and recently, under the young historian Giovanni Spadolini as editor, it has really distinguished itself. Galli in his book, *Il Bipartitismo Imperfetto*, attaches great importance to Mattei's *Il Giorno* which comes out in Milan and has a very modern 'face', yet one never sees anyone reading it. On the whole the Italians are still quite extraordinarily traditional about their newspaper-reading. The Communists read *Unità*, most Socialists *Avanti!* and most Social Democrats *Umanità*, but all the Milanese appear to read the *Corriere*, all Torinesi *La Stampa*, and all Venetians *Il Gazzettino* founded in 1886. There is rather more variety in Rome with *Il Messagero* (founded 1879), *Il Giornale d' Italia* (founded 1900), and of course the *Osservatore Romano* from the Vatican. It is not uninteresting that the one-time Secretary-General of the Christian Democrat Party, Piccoli, owns the *Adige* newspaper at Trent. There is a wealth of periodicals, among which the best-known are *Il Mondo* which constantly dies and is reborn, *Il Mulino* at Bologna, *L'Epoca*, *L'Espresso* and so on. In the fifties

Silone started *Tempo Presente* together with Nicola Chiaromonte, expert on the theatre; this has already been referred to, as has also the most remarkable of all Italian periodicals, *Il Verri*, named after Pietro Verri, the *Illuminato*. It is published in Milan by Feltrinelli every two or three months; its editor is Luciano Anceschi. Its thirty-first number came out in December 1969, full of fascinating studies: a leading article by the editor on Kant, several articles dealing with T. S. Eliot and one with Joyce. It also contained an article on the poems of the painter Kokoschka and one on the English sense of humour. In no. 32 (March 1970) Anceschi and others wrote on Jean Paulhan. No. 33 dealt with Lucini and Futurism and thus of course with Marinetti. *The Times Literary Supplement* (12 March 1971) reviewed this number. It said: 'Luciano Anceschi deserves our warmest congratulations for being able to keep alive *Il Verri*, one of the best and most thoughtful literary reviews now existing. It is international in spirit and, in its time, has drawn on many of the best-known critics from the late Renato Poggioli to F. R. Leavis.'

Since Ungaretti's death in 1970 at the age of eighty-two the doyen of Italian poets is Eugenio Montale who was born in Genoa in 1896. Bruised by Fascism into the leadership of the 'hermetic' school, he emerged in 1956 with his collected poems published under the title of *La Bufera e altro*. He is the least rhetorical of poets. The outstanding poet of that decade was, however, the Sicilian Salvatore Quasimodo, sad and hermetic too and only five years younger. His poem *La Terra impareggiabile* in 1958 brought him a Nobel prize for literature in 1959. In the next generation, men born on the eve of the First World War, Vittorio Sereni, Alfonso Gatto and Mario Luzi, are distinguished. They are all much influenced by the war of 1940 and the *Resistenza*. One feels the influence of Rilke and, in the case of Luzi, that of T. S. Eliot: these Italians are poets of suffering. One of the youngest of the poets of Italy to have made a name is Pier Paolo Pasolini, born in Bologna in 1922. This strange, tortured genius, often said to be unidentifiable, is better known as the producer of famous films. But he is probably equally important as a poet, famous poems of his including *Il canto popolare* (1954), *Ragazzi di vita* (1955), *Canzoniere italiano* (1955), *Le ceneri di Gramsci* (1957). He is the archetypal figure of the Christian–Communist dialogue (see Chapter 8 above), for he is torn between

Christ and Chairman Mao. He is a kind of twentieth-century Marlowe, precocious and immature at the same time, fascinated by blasphemy and cruelty, but also by virtue which he seldom finds except among 'the people'. The poet Edoardo Sanguineti is eight years younger; he is an essayist as well and an authority on Dante.

The years since 1945 have been a glorious period for the Italian cinema. Beginning with Rossellini's *Roma Città aperta* with Anna Magnani playing in it, through the films of De Sica and Antonioni, they reached what many regarded as a climax in Fellini's *La Dolce Vita* – *la douceur de vivre* brought up to date – in 1959 and *Otto e mezzo* in 1962. In 1970 he produced *Satiricon*. Francesco Rosi and Ermanno Olmi were also notable film producers. The fluency and brilliance of improvisation natural to Italians contributed to the success of their films; the vibrating vitality of Italian crowds provided one kind of inspiration.

Two other film producers with strong leftist sympathies are the poet Pasolini and Luchino Visconti, Duke of Modrone. Pasolini began by acting for Fellini and then in Carlo Lizzani's film *Il Gobbo*. His own first film was *Accattone* in 1961, a brilliant study in squalor. Later he produced the beautiful *Gospel according to St Matthew*, and then again later (1969) the terrible *Porcile* or pigsty. (Anna Magnani played in his second film, *Mamma Roma*, in 1962.) Visconti, who like so many others began with Jean Renoir, in *Rocco e i suoi fratelli* examined a southern peasant family faced with Milan. Visconti was equally successful in the splendid film he made of the *Gattopardo* in 1963 and in the production of opera and drama; his name has often been linked with that of the gifted Franco Zeffirelli. Now Visconti has produced the film of Thomas Mann's novella, *Death in Venice*, changing the central figure into a musician based on Mahler whose music is used. This was made as part of the campaign to save Venice from being submerged by the sea, and if it seems faintly perverse to have chosen to portray the city at a time of pestilence, yet the film is of the purest beauty and historically meticulous, a very mature work of art. After this Visconti, who has previously interested himself in Thomas Mann, intends to make a film of the middle volumes of *A la recherche du temps perdu*. One can predict with confidence that he will do no wrong to Proust – can more be said?

It would not be wise to try to list Italy's film stars in this period, but perhaps Marcello Mastroianni (born 1923) is so outstanding as to demand a mention. There is no Italian in the theatre whose name can compete with his or with that of the actor, Ruggero Ruggeri, of an earlier generation.

The theatre had hitherto been dwarfed by the opera in Italy, but after 1950 many more regular companies were formed. Beside the majesty of the opera-house of La Scala, often mobbed recently by young leftists on first nights, the *Piccolo Teatro* was established in Milan. Under the direction of a young Triestine, Giorgio Strehler, who had begun in French Switzerland during the war, the *Piccolo* produced all the best plays one could think of. The extraordinary thing was that the new Italy produced no playwright; the only Italian name on the list was that of the Sicilian, Pirandello, who had died in 1936. It was the same at the *Teatro Stabile* at Genoa though it had no Strehler. Much of the acting was magnificent. Albertazzi's *Hamlet* produced by Zeffirelli may be said to have competed with the best Hamlets of London or Stratford. While providing no playwright, the new Italy has an admirable dramatic critic in Nicola Chiaromonte, already referred to as co-editor of *Tempo Presente*. In 1959 Chiaromonte published his only book so far, entitled *La situazione drammatica*. In this he wrote:

> Truth and freedom are the watchwords of modern drama. . . .
> And . . . if the theatre in Italy, though full of talent and of
> potentialities, has remained confused and uncertain up to now,
> and is still – despite the great example of Pirandello – not
> truly modern, it is, essentially, because the spirit of truth and
> freedom is still cramped and impeded in our theatre . . . and,
> above all, because it has not really been understood.

This was three years before the Vatican Council was to meet and in a period when such plays as Hochhuth's *Stellvertreter* were blocked by the hierarchy, Ottaviani and company. Now in 1971 it is different, but the Italian hierarchy lags behind the spirit of the times still, and it is difficult to be sure what Pope Paul VI wants, for it is not clear that he himself knows.

It seems almost superfluous to mention that Pasolini published a play called *Orgia* in 1969 – it has not inspired much comment. It should also be recorded that Moravia together with Pasolini and Alberto Carocci bring out an influential review called *Nuovi Argomenti* – this has been going since 1953.

The leading Italian composers of music since 1945 have been Luigi Dallapiccola and Goffredo Petrassi. Toscanini's return from America at the end of the war was a tremendous occasion; he died in 1957, ninety years old. Other great conductors have been De Sabata, Gui and Giulini, while as a pianist Michelangeli is famous, *The Times* describing him as a 'princely' performer when he visited London in June 1965. Luigi Nono, Bruno Maderna and Luciano Berio should also be mentioned as performers. The leading painters have been De Pisis, Morandi and Guttuso, and the sculptors Manzù and Marino Marini are deservedly well known. In architecture Nervi became more famous after the construction of his Exhibition building in Turin and after his contribution to the United Nations building in Paris. He also contributed to the Pirelli building in Milan, which was, however, essentially the work of Gio Ponti; three younger Italian architects who have been doing well are Rogers – he is Italian in spite of his English name – of the Studio BBPR, Franco Albini and Gino Valle – and there are many others. Rogers together with Belgioso and Peresutti redesigned the Castello Sforzesco in Milan. Gio Ponti has joined in the Italians' success in designing furniture. Adalberto Dal Lago, Alfredo Pizzogreco and Sergio Asti have also distinguished themselves in this field. Finally the Italian Republic has overtaken the French in dress design with an astonishing degree of variety, verve and stylishness.

Conclusion

The life of ordinary people in Italy has changed tremendously since 1945. Of course it has everywhere, but Italy is an extraordinary mixture of developed, developing and undeveloped. This probably means that there has been more change in the life of more individuals than anywhere else one can think of. The Russian, whose grandparents, certainly, were illiterate, has learned to read and write, but he is not free to spend his money because so much is deliberately kept out of his reach. The Italian is only hampered by his inefficient bureaucracy. Wages have risen very appreciably since the hot autumn of 1969. Indeed the chief threat to Italian prosperity is the continued striking which has reduced production in 1970 and could seriously threaten the newly acquired standard of living.

Enough has been said of the *Mezzogiorno*, perhaps not enough of northern Italy. Milan has a sophisticated stylishness of its own so that within the city one can remain unaware of the immigrant southern workers in the suburbs. If one takes one of those splendid trains to Venice one is almost certain to meet a Venetian who now lives in Milan; he is going back to Venice to visit his parents but is thankful he no longer lives there as it is a dead city. When one gets to Venice life on the *vaporetti* seems very vigorous, and recently I have seen prams labouring up and down the bridges as I never had before. Dazzled by the beauty of the buildings of Venice, one cannot, nevertheless, fail to notice that the water is a little higher each year near the Doge's Palace. The smaller towns of the *terraferma* live on in their delightful way, the grip of the priests loosening a little as the years pass. Turin and Genoa on the other side of northern Italy are beautiful in other ways, and lake-life on Maggiore, Como or Garda is both unique and lovely.

It has been seen that central Italy may be considered as the best-organised portion of the country. Many would regard it as the most Italian. The peasants of Emilia-Romagna and Tuscany have been able to go into local industries which have not swollen to inhuman proportions. That glorious Tuscan landscape makes one's heart leap up, but it sinks again when one hears that the cypress trees have a strange disease – probably due to some kind of pollution – which may prove fatal to them.

Nevertheless in 1970 rising fortunes made the dominant impression. A typical example was that of a postmistress in Terontola, the railway junction on the frontier between Tuscany and Umbria almost on Lake Trasimeno. She was in her forties, coming from a Tuscan peasant family. She had kept a general store and made large sums during the years of the miracle. By 1970 she had built herself a pleasant house and owned two cars; her husband was a postman too and they delivered the letters by car. She felt that petrol was very expensive after the *decretone* but she had voted for the Christian Democrats in June and seemed likely to do so again. The local authorities in her area were, of course, mostly Communist. Her family was large and she had nephews who were schoolmasters. The mentor of the large family was a well-known aristocrat for whom some of them worked domestically. Their father, they told me, had died happy because he knew that the Count would always look after them. But the Count has grown old, and the times, in a sense, less indulgent, although perhaps more just; in the future these people must fend for themselves.

Select Bibliography

GENERAL

Carlyle, Margaret, *The Awakening of Southern Italy* (1962)
D'Alfonso, A. (ed.), *I Cattolici e il Dissenso* (1969)
De Rosa, Giuseppe, S. I., *Cattolici e Comunisti oggi in Italia* (1966)
Falconi, Carlo, *Pope John and his Council*, trans. M. Grindrod (1964)
Galli, Giorgio, *Il Bipartitismo Imperfetto* (1966)
Ghirotti, Gigi, *Mariano Rumor* (1970)
Grindrod, Muriel, *The Rebuilding of Italy* (1955)
——, *Italy* (1968)
Hughes, H. S., *The United States and Italy* (1965)
Kogan, Norman, *A Political History of Post-war Italy* (1966)
Lucarini, Spartaco, *Democrazia in crisi* (1970)
Mack Smith, Denis, *Modern Sicily* (1968)
Nenni, Pietro, *Le Prospettive del socialismo dopo la destalinizzazione* (1962)
Pallenberg, C., *Vatican Finances* (1971)
Spriano, Paolo, *Storia del Partito Comunista Italiano*, vol. 1 (1967); vol. 2 (1969)
Trevelyan, Raleigh (ed.), *Italian Short Stories* (1965)
Webster, Richard, *Christian Democracy in Italy, 1860 to 1960* (1961)
Wiskemann, Elizabeth, *Italy* (1947)

ECONOMIC

Amendola, Giorgio, *Classe operaia e programmazione democratica* (1966)
Annuario Statistico, yearly
Clough, S., *The Economic History of Modern Italy* (1964)

La Malfa, Ugo, *La Politica economica in Italia* (1962)
——, *Verso la Politica di Piano* (1963)
La Palombara, J., *Italy: The Politics of Planning* (1966)
Lutz, Vera, *Italy: A Study in Economic Development* (1962)
Rossi Doria, Manlio, *Riforma agraria e azione meridionalista* (1948)
—— *Dicci Anni di politica agraria nel Mezzogiorno* (1958)
Saraceno, Pasquale, *Iniziativa privata e'azione pubblica nei piani di sviluppo economico* (1959)
Sylos-Labini, Paolo, *Problemi dello sviluppo economico* (1970)

Chronology of Chief Events, 1945–1970

1945	June	First free Government headed by Ferruccio Parri
	Sep	*Consulta* meets
	Dec	Alcide De Gasperi's first Government
1946	Feb	Dissolution of Party of Action
	2 June	Referendum in favour of a republic: House of Savoy expelled
		Constituent Assembly elected
		De Gasperi–Gruber agreement on South Tirol
1947	Jan	Saragat breaks away from Nenni's Socialist Party and founds own Social Democratic group
	May	Nenni and Togliatti leave the Government. Sforza succeeds Nenni as Foreign Minister
	Aug	Drastic credit restrictions
1948	1 Jan	New constitution comes into force
	Mar	Western Allies declare for return of Trieste to Italy
	Apr	General election; Christian Democrat success
	May	Luigi Einaudi elected first President of the Republic
1949		Discovery of oil and gas in Po valley
1950		Agrarian reform introduced
1951		Amintore Fanfani Minister of Agriculture
		Death of Sforza
1952		First national congress of neo-Fascist *Movimento Sociale Italiano* (MSI)
1953	Feb	Formation of ENI
		General election; Christian Democrat losses, Communist gains
		De Gasperi retires

		Oil discovered in Sicily
1954		Death of De Gasperi
	Oct	Italo-Yugoslav agreement on Trieste
1955		Gronchi succeeds Einaudi as President
		Antonio Segni Prime Minister
		Italy joins the United Nations
1956		Economic miracle begins
		Italy gets commission to build the Kariba dam
1957		Mattei's agreements with Egypt and Iran
		Segni resigns
1958	May	General election
		EEC Treaty signed in Rome
		Fanfani Prime Minister
		Death of Pope Pius XII; Cardinal Roncalli elected Pope John XXIII
1959	Jan	Fanfani resigns as Prime Minister and Secretary General of the Christian Democrat Party
		Aldo Moro succeeds as Secretary-General
	Apr	The Holy Office condemns political collaboration with all Marxists
1960		Tambroni's rightist bid for power fails
1961	Apr	Pope John publishes Encyclical *Mater et Magistra*
		Centre-Left town councils in Milan and Genoa, Centre-Left Government in Sicily
	Sep	Pasquale Saraceno demands economic planning at Christian Democrat convention
1962	Feb	Fanfani's planning Government in power
		Ugo La Malfa Minister of the Budget and of Planning
	Sep	Second Vatican Council meets for first session
	Oct	Death of Mattei
		Segni President of the Republic
	Nov	Electricity nationalized
1963		Economic recession sets in
	10 Apr	Encyclical *Pacem in terris* appeals to 'all men of goodwill'
		General elections; Communist gains
	June	Death of Pope John
		Cardinal Montini elected Pope Paul VI

1963	Sep–Dec	Second session of Vatican Council
	Dec	First Centre-Left Government under Moro, includes Nenni's Socialists and Nenni as Deputy Premier
		PSIUP breaks away from Nenni
1964		First Five-Year Plan announced
	Mar	Ippolito arrested
		Rumor elected Christian Democrat Secretary-General
	July	Government crisis; talk of dictatorial plans of Pacciardi and De Lorenzo
	Aug	Segni incapacitated by stroke
		Death of Togliatti
		Third session of Vatican Council
	Dec	Saragat elected President
		Fanfani succeeds him as Foreign Minister
1965		Economic recovery
		Law for abolition of the *mezzadria* system
		Final session of Vatican Council
	Sep	Fanfani elected President of the UN Assembly
	Nov	Conference to discuss reunification between PSI and Social Democrats
1966	Jan	Eleventh Congress of the PCI
	Aug	Collapse of new houses at Agrigento
	30 Oct	PSI unites with Social Democrats
	Nov	Floods in Tuscany and Venetia
1967		Student unrest becomes marked
		Fortuna introduces divorce bill into Chamber
	Apr	Dismissal of De Lorenzo as Chief of Staff
1968		Regions voted into existence
	Feb	Meetings of *Gruppi di impegno politico* to protest against conservatism of the Church and in favour of a *Repubblica Conciliare* based on Catholic–Communist co-operation
	May	General Election; Socialist failure
		Moro resigns
		Leone heads caretaker Government
	Aug	PCI condemns Russian invasion of Czechoslovakia

	Dec	Rumor Prime Minister, Nenni Foreign Minister
1969	May	Divorce proposed again in Chamber
	July	Nenni retires
		Partito Socialista Unitario founded by Tanassi as opposed to PSI
	Aug	Rumor's second Government, not *Centro-Sinistra* but Christian Democrat only
1969		Donat Cattin Minister of Labour
		Amendola demands that PCI be included in Government
	Sep	Revolution in Libya
		'Hot autumn' of strikes begins
	12 Dec	Bomb explosions in Milan and Rome
	21 Dec	New contract signed by employers and metal-workers
1970	Feb	Pope condemns divorce
	Mar	Rumor forms new Centre-Left Government with Moro at Foreign Office
		Strikes in favour of social reforms
	14 May	Labour Charter passed
	7–8 June	Elections in many communes for the communal provincial and new regional assemblies; Centre-Left Government confirmed
	July	Rumor resigns
		New Fiat contract
	Aug	Colombo Prime Minister
	Dec	Chambers agree to revised Economic Decree and to divorce

Index